MINING IN THE OUTER CONTINENTAL SHELF AND IN THE DEEP OCEAN

Panel on
Operational Safety in Marine Mining

MARINE BOARD
ASSEMBLY OF ENGINEERING
NATIONAL RESEARCH COUNCIL

NATIONAL ACADEMY OF SCIENCES
Washington, D.C. 1975

NOTICE

This is the report of an activity undertaken with the approval of the Governing Board of the National Research Council, representing and consisting of members of the National Academy of Sciences, National Academy of Engineering, and Institute of Medicine. Such approval signifies that the project is of national importance and appropriate to the purposes and resources of the National Research Council.

The members of the panel appointed to undertake the project were selected for recognized competence and with due consideration for the balance of disciplines and expertise appropriate to the activity. Responsibility for the substantive aspects of this report rests with that panel.

Each report by a panel of the National Research Council is reviewed by an independent group of qualified individuals according to procedures established and monitored by the National Academy of Sciences, National Academy of Engineering, and Institute of Medicine. Only upon satisfactory completion of the review is the report approved for publication.

ISBN 0-309-02405-6
Library of Congress Card Catalog Number 75-24928

This report represents work supported by Contract 14-08-0001-14378 between the U.S. Geological Survey and the National Academy of Sciences.

Available From:

Printing and Publishing Office
National Academy of Sciences
2101 Constitution Avenue, N.W.
Washington, D.C. 20418

Printed in the United States of America

PREFACE

As the worldwide demands for basic minerals increase and some resources on land show signs of rapid decrease, it is inevitable that the search for new sources will extend to the oceans--the largely unexplored 71 percent of the planet's surface. With modern technology, the ocean depths are already tapped. At present, oil extracted from the ocean floor supplies about 19 percent of the world's need. By 1980 offshore production, it is estimated, will account for between 30 percent and 40 percent of all the oil and perhaps 10 percent of the natural gas. Of possibly equal importance is the rich reserve of hard minerals in the oceans--for example the so-called manganese nodules, which vary in size and shape from small pebbles to massive pavements, containing economically attractive deposits of manganese, copper, cobalt, and nickel.

It is important to assess the nature and extent of the mineral resources of the seabed and to devise ways of recovering them with a minimal impact on the environment.

This report attempts to do that--to examine the potential of the resource, evaluate the state of the art of ocean mining, identify the legal, regulatory, and jurisdictional problems involved, consider the possible environmental questions, and determine how to meet the needs for trained engineers to do the job.

The report, based on an 18-month study, is especially concerned with mining hard minerals in the outer continental shelf and in the deep ocean plains. It does not deal with the production of oil and natural gas, the mining operations in the coastal zone, the recovery of salt and sulfur from the continental margins, or such matters as miner safety, metals processing and national security.

Although it was not mentioned in the report, it was the consensus of those panel members most concerned with environmental protection that the development of mining surveys and ecosystem monitoring begin with a prototype mining operation.

The panel was deeply saddened during its deliberations by the death of one of its members, the eminent metallurgist, Antoine Gaudin.

J. Robert Moore, Chairman
Panel on Operational Safety
in Marine Mining

SUMMARY

Over the past 25 years, the United States has grown increasingly disturbed about its potential vulnerability to the supply of minerals in the world and the actions of other nations possessing those resources. In the early 1950's, President Truman, concerned about the acute shortages experienced during and after World War II and the Korean War, established the Paley Commission to explore the problem.[1] Major studies by the National Academy of Sciences, the National Academy of Engineering, and others, have called attention to the steadily declining quantities of certain material resources and the close relationship between availability, production and application as well as such concerns as environmental protection, foreign supplies, world competition and national goals.[2] In the most recent of those studies, Mineral Resources and the Environment, a committee of the National Research Council, the operating agency of the academies, issued a forceful statement on the problem:

> "Copper resources in manganese nodules on the floors of the deep seas are apparently as large as developed reserves in conventional deposits on land. Because there is uncertainty as to the ability of the United States to meet demands for copper from domestic sources, we recommend that developing the recovery of copper and associated metals from these nodules be encouraged with due regard to the potential impact of undersea mining of the environment.[3]

In this report by a panel of the Marine Board, Assembly of Engineering, of the National Research Council, the state-of-the-art of mining exploration and mineral recovery on the outer continental shelf and abyssal oceans plains is evaluated. Assessments are made about the potential availability of hard minerals, with estimates of quantity, feasibility, and value--the criteria that are always applied in making decisions to go ahead or not. In the case of copper, cobalt, manganese, and nickel, which are present in manganese nodules, there are additional factors, principally their importance to an industrialized society like that of the United States.

Although the emphasis of the study was on technological factors, the panel examined such important related issues as regulatory and leasing requirements, environmental implications, and education and manpower needs. Drilling for oil and natural gas was not a subject for the panel's analysis--nor were such matters as mining safety, metals processing, and national security.

Complex, difficult and costly as it is bound to be, marine mining offers enormous potential for becoming independent of foreign countries for some important minerals, including those used as sources of energy. The United States is fortunately located near several prime and secondary sites for these minerals-- principally on the continental shelf north of Virginia, the Gulf of Mexico, California and Alaska. Estimates of the abundance and worth of these resources are contained in Table 3 (pages 8-11).

TABLE 1. Marine Mining Categories.

Ores	Operating Depth[1]	Probable Mining Rate[2]	Probable Mining Equipment[3]	Status of Mining System Development[4]
Sand & gravel	Shallow	High	Hopper	Complete, U.S./Foreign
			Hyd. C-H	Complete, Foreign
			Mech.	Partial, Foreign
Heavy minerals	Shallow	High	Mech.	Complete, Foreign
			Hyd. C-H	Partial, Foreign
Cassiterite	Shallow	High	Hyd. C-H	Complete, Foreign
			Mech.	Partial, Foreign
Diamonds	Shallow	Low & high	Pneumatic	Complete, Foreign
			Hyd. C-H	Complete, Foreign
Gold	Shallow	High	Mech.	Complete, Foreign
			Hyd. C-H	Partial, Foreign
Barite	Shallow	Low	Mech.	Complete, U.S.
Phosphates	Shallow & intermediate	Low & high	Hyd. C-H	Partial, Foreign
			Mech.	Partial, U.S./Foreign
Manganese nodules	Deep	Low	Pneumatic	Partial, U.S./Foreign
			Mech.	Partial, U.S.
			Hyd.	Partial, U.S.

1. Shallow = 150 ft. or less; Intermediate = 150 ft. to 3000 ft.; Deep = deeper than 3000 ft.
2. Low = 500 cubic yards/hour or less; High = more than 500/hour.
3. Hyd. C-H = Hydraulic cutterhead; Mech. = Bucket Lift; Pneumatic = Air Lift System.
4. Sea state compensation development only may be required where indicated "Partial," except for manganese nodules. U.S. or Foreign refers to location of operations except for manganese nodules where it indicates technology development.

On the continental shelf, the Panel believes that initial mining operations in the production of sand and gravel will continue to be conducted with rather conventional equipment (Table 1). On a smaller scale, and with similar conventional equipment, other resources, such as rare earth sands, barite, coal, tin, and phosphate rock have already been produced from shelf deposits in various parts of the world. Such activities are expected to increase as technological capability and economic rewards increase. Unlike the area underlying the deep ocean, the question of ownership of much of the continental margins of the world is well-defined under existing international law.

In the deep ocean, the situation is much more complex, although the early production of basic resources seems much more promising. This is due primarily to the extensive deposits of ferromanganese nodules found throughout the world ocean seafloor at depths of 3600-5500m (12,000-18,000 ft). In many deposits these nodules contain high grade ore-quality manganese, copper, cobalt and nickel. While the development of the technology for this type of mining is considered well in hand, the initial capital investments are very high. Nevertheless, at least four corporations or consortia are now planning to initiate deep sea mining operations on commercial scales. A major inhibiting factor is the disputed international law of the sea regarding who has the right to mine deep ocean mineral resources and under what conditions? Whether or not this question is resolved in each case by the United Nations, the United States government is expected to legislate sufficient guarantees to permit initiation of commercial deep-ocean mining to begin within the next two to four years. Once this happens, mining other marine resources from the deep seafloor should expand.

While the Panel recognizes the importance of processing nodules for the particular metals of value, the details and methodology of extractive metallurgy are topics beyond the scope of this report. The state of technology today indicates that processing will be initially carried out on land, using technology especially developed for the nodules. Processing at sea using these techniques would require platforms with little or no motion and space often exceeding 20 hectares (50 acres).

With the initiation of marine mining operations, careful studies and assessments of their environmental impact will be required. The mandate to accomplish this on the continental shelf presently exists through the National Environmental Policy Act (NEPA), which established the basis for a regulatory framework for all activities that might have environmental impacts. In the international region of the deep ocean, the case is not clear. However, it has been ruled that an environmental impact statement will be required before any United States involvement can be legislated, and it is reasonable to assume that the government will impose environmental guidelines on the United States firms working in this area. While the problems of deep-ocean work are being resolved, there is an opportunity to carry out the environmental assessment before any commercial activity is undertaken. This was not the case with mining on the continental shelf, which in some cases preceded any environmental rules.

The regulation and leasing requirements are simpler on the continental shelf, because the title to the lands involved is relatively clear in most places. The regulation and leasing arrangements will invoke questions of environmental protection, offshore safety of operations, interference with other uses of the continental shelf waters, concessions to proven exploitation capabilities, and revenues to state and federal governments. In the deep ocean these same questions must be addressed, although to differing degrees. A principal problem is the determination of the rights of ownership. Past experiences of

United States regulatory and leasing agencies should serve as a useful guide--both for how and how not--for establishing incentives for attracting capital investment to the new venture of ocean mining.

The requirement for trained manpower in the various component areas in marine mining, from management to technology and from marketing to field work, is not now considered a major limiting factor. However, the lag between the establishment of curricula and the hiring of the first graduates makes it prudent to consider now what might be the future needs of the marine mining industry. College-level training will be needed at the undergraduate and graduate level in related sciences, engineering, and management. Technical school curricula will be required to train shipboard operators and technicians to handle the mining platform and its equipment. An important phase of the entire process of education is the stimulation of public awareness of marine mining. Without public knowledge, interest, and support, the initiation of commercial marine mining will be more difficult.

In the opinion of the Panel, adequate educational facilities now exist to support the estimated needs of a marine mining industry. What is required in these institutions is some reorganizing of the existing capability to meet the specific needs of this new field.

SPECIFIC FINDINGS

As a result of its deliberations, the Panel offers the following conclusions and recommendations:

Outer Continental Shelf	Deep Ocean
Importance and Potential	
Conclusion. The development of mineral resources has positive economic potential and it appears that mining can be conducted within acceptable limits of environmental risk (as weighed against the expected economic gain).	Conclusion. Same
Recommendations. *The federal government should take steps to (1) provide incentives for the development of mineral resources by the establishment of appropriate regulations and leasing procedures; (2) undertake a continuing assessment of*	Recommendations. *Same*

Outer Continental Shelf — Deep Ocean

Importance and Potential (cont'd)

the mineral resources, and (3) support development of exploration and interpretive technology as items of first priority.

Mining Technology

Outer Continental Shelf

Conclusions. Although economic criteria must be considered, the technology now exists to support outer continental shelf mining of unconsolidated hard mineral deposits by dredging, in water depths to 92 meters (300 ft). These depths can be doubled with minimal development. Mining of consolidated hard-mineral deposits is also feasible in some areas of the outer continental shelf by open-pit excavation or by means of shafts sunk either on land or through artificial islands. Further, resources amenable to fluid extraction may be recovered through drill holes sunk from fixed or floating platforms in any area of the outer continental shelf.

Outer continental shelf mining of certain mineral deposits appears economically viable; however, the cost of leasing and of meeting still undefined environmental regulations could affect this conclusion.

Outer continental shelf mining differs significantly from terrestrial mining, deep-ocean mining, and offshore petroleum operations.

Recommendations. *Given the above, the Panel recommends*

Deep Ocean

Conclusions. Exploration techniques used by government, university, and industry researchers have thus far been adequate to identify broadly the potential of ferromanganese nodules found in the oceans; however, present exploration techniques need to be improved to meet the future demands of full-scale mining operations. Equipment and subsystems appear to be available for deep ocean mining systems, but no total system has yet been proven reliable to support commercial operations.

Four areas of engineering knowledge pertinent to marine mining require a considerable amount of improvement. These are:

(1) fatigue life of materials in seawater;

(2) fracture mechanics of materials exposed to seawater and other corrosive media at high stress levels;

(3) the effect of residual stresses due to welding; and

(4) engineering properties of marine sediments.

Recommendations. *With regard to materials and structures*

Outer Continental Shelf | Deep Ocean

Mining Technology (cont'd)

that (1) the Department of the Interior recognize the significant differences between outer continental shelf mining and deep-ocean mining, offshore petroleum operations and land mining before establishing final outer continental shelf hard mineral regulations; (2) one or more heavily monitored, commercial scale, prototype operation(s) should be sponsored by the federal government for the dual purposes of generating design criteria for the fabrication of environmentally sound mining systems and devising environmentally sound leasing regulations and operating procedures; (3) private enterprise be assured of lease terms of sufficient duration and area to allow for amortization of the major investment required for mining equipment; and (4) government-sponsored research and development of value to ocean engineering be continued in relevant areas of the marine environment, on navigation techniques, on structural materials in seawater, on effects of weather, and on marine geology and soil mechanics.

for marine mining applications, the Panel recommends that the federal government and industry perform a comprehensive materials testing program in order to evaluate selected materials and characteristics of those materials, complete the program by 1978, and publish a report for public use. Independent laboratories are recommended as the test agencies. The testing program should consider:

1. *Materials: steels, titanium, aluminum;*
2. *Form of Test Specimens: weldments, forgings, sheet, and plate;*
3. *Evaluation Parameters: mean stress level, stress ratio, S/N curve* in air and water, and stress concentration factors.*

With regard to component reliability, the Panel recommends that (1) a joint government and industry program be organized to form standards similar to the approach of Det norske Veritas.[4] Components that are common to several engineering problems such as pressure compensation should be standardized. Others that are custom-designed should follow standardized development guidelines, and acceptance testing criteria. The Panel further recommends that (2) guidelines be

* S/N = Stress Level/Number of Cycles

established, based on the comprehensive materials testing program, for determining structural design criteria of deep-ocean systems; and that

(3) the United States Government take the lead by supporting industry in the development of sensors and survey systems unique to ocean mining, such as in-situ mineral content analyses of ferromanganese nodules and microbathymetry survey systems. Basic research in this field, when conducted by or for the government , should be made public.

For the collection of environmental data, the Panel recommends that (4) the United States government, to help meet the needs of operational safety in marine mining, produce a sea-state prediction model, augmented by suitable buoys and sensors, for areas of ocean mining interest (0° - 20° north latitude, 120° - 180° west longitude). The Panel further recommends that (5) a national clearinghouse for collection of soil mechanics data be established and that industry cooperate in the formation of maps containing geological and geophysical data. These data should be collected and submitted for expert analysis to determine if additional work is required. Upon completion of analysis, the data should be published by the United States government.

Environmental Protection and Safety

Conclusions. There will be environmental impacts associated with the onshore activities that accompany offshore mining. Some will be associated with the transport of the minerals (marine terminals and support facilities, stockpiling of materials, truck movements, etc.) and others will be associated with the processing of the minerals. Assessment of the environmental impacts of these activities should proceed before there is a move to license full-scale offshore mining. To assure development of environmentally safe ocean mining, industry must be willing to disclose data on the technology of mining pertaining to those elements of mining systems that directly interact with the environment. Ideally, such requirements would be satisfied by providing, without restriction, the functional information and specifications, rather than the detailed design or process data. However, recipient groups must be willing to receive and maintain any proprietary information under the terms of protective confidential disclosure arrangements that prevent public access to the data. It must be recognized that breach of such agreements and the resultant compromise of proprietary information could result in serious retardation of the efforts to make the benefits of ocean resources available for world use. With the exception of those proprietary data disclosed in confidence, the output from this cooperative endeavor of government, academia, industry and other interested groups related to the environmental impact of

Conclusions. Same

Outer Continental Shelf	Deep Ocean

Environmental Protection and Safety (cont'd)

ocean mining should be subject to public scrutiny.

It is essential for the protection of the environment to develop orderly procedures for mining on the outer continental shelf. In light of its present stage of development, the outer continental shelf mining industry has an opportunity to design hardware and operating methods that will minimize potential adverse environmental consequences.

The extent of environmental information presently available on the outer continental shelf varies, with some areas such as in the Gulf of Mexico relatively well studied and others, say, off Alaska, relatively unknown. Until such knowledge and understanding is obtained, extrapolation of the available data and information from one area to another will be difficult.

Standard criteria do not exist for gathering and evaluating data describing the environment for a given outer continental shelf area and for a given scope of activity. Proposed standards are currently being developed by a number of agencies. The environmental impact of specific mining operations, as well, is only partially understood because to date there has been little experience in mining of the outer continental shelf.

Environmental safeguards for outer continental shelf mining could be maintained if (1) adequate prelease base-

Although our knowledge of the impact of manganese nodule mining on the oceanic environment is developing and the environmental effects of two mining tests have already been monitored,[5] [6] it remains difficult to forecast precisely what the environmental effects of full-scale mining operations will be.

A unique opportunity presently exists to further the close collaboration between the government, the marine mining industry, and other interested groups through joint programs to develop plans for environmentally-safe deep ocean mining before commercial operations start.

Outer Continental Shelf | Deep Ocean

Environmental Protection and Safety (cont'd)

line studies are performed; (2) provisions for environmental monitoring are stipulated; and (3) provisions for evaluation and corrective action are mandated in the regulatory system. Foreign continental shelf mining experience can be of value in predicting certain environmental problems to be expected on the outer continental shelf.

Recommendations. *The Panel recommends that (1) cooperative industrial and governmental research to identify the existing environmental conditions in potential outer continental shelf mining areas be intensified, and further, that research be conducted to identify the environmental effects, both short- and long-term, of outer continental shelf mining; (2) prototype operations be undertaken in representative areas (environmental effects of these prototype operations should be monitored so that the long-term impacts can be weighed against the short-term changes before full-scale leasing is begun); (3) environmental standards be formulated using the knowledge gained during monitoring of prototype operations; (4) industry be informed of pertinent environmental requirements in a timely manner through the formulation of sound regulations and environmental criteria; (5) an independent panel of experts be established to provide for adequate review of proposed scopes of work, and for evaluation of*

Recommendations. *The Panel recommends that appropriate agencies of the United States government be responsible for (1) determining baseline environmental conditions in the potential mining areas (to be continued and completed, if necessary, during the subsequent phases of the procedure); (2) environmental monitoring of pilot and/or full scale mining operations; (3) documentation of changes induced in benthic and pelagic ecosystems by deep-ocean mining and evaluation of their implications in relation to current and potential marine resources; and (4) if necessary, recommendation of changes in mining methods and equipment based on the facts established in 2 and 3 above.*

Based on the findings described in 1 above, the United States Government should (1) prepare an adequate programmatic environmental impact statement for manganese nodule mining; (2) formulate environmental criteria and regulations for the mining operations to minimize possible harmful environ-

Outer Continental Shelf — Deep Ocean

Environmental Protection and Safety (cont'd)

Outer Continental Shelf

results of baseline and monitoring studies, and (6) relevant foreign continental shelf mining operations be evaluated with respect to their impacts on the environment.

Deep Ocean

mental effects while enhancing the development of potentially beneficial effects; (3) evaluate the environmental impact reports submitted by the mining companies in support of their applications for lease and production licenses, where required; (4) prepare the specific environmental impact statement for each lease and production license, where required; and (5) monitor and enforce the environmental regulations until an international agreement can be reached.

Outer Continental Shelf — Deep Ocean

Regulations and Leasing

Outer Continental Shelf

Conclusions. There is at present a very limited experience base upon which to build a licensing and regulatory system for hard-minerals mining in the oceans; there is a need for a regulatory licensing system that is flexible and can be continually upgraded as a result of lessons learned from early mining activities; and that the Outer Continental Shelf Lands Act has major deficiencies as a legislative basis for leasing and regulation of hard-minerals mining.

Deep Ocean

Conclusions. Same.

Outer Continental Shelf

Recommendations. *Generally, the Panel recommends an approach to government regulation and management of outer continental shelf mining that replaces early financial advantage and administratively clean*

Deep Ocean

Recommendations. *The Panel recommends that (1) prospecting, using such methods as magnetic, gravimetric, and seismic surveys, as well as bottom sampling and shallow coring, may be undertaken by any United States citizen or*

Outer Continental Shelf | Deep Ocean

Regulations and Leasing (cont'd)

allocation of mining leases by sealed bids, based on bonuses, with the early and complete information advantages of a licensing system which uses work program proposals as a basis for allocation.

Specifically, the Panel recommends that (1) prospecting be available to any United States citizen or company upon issuance of a permit by the Department of the Interior; (2) the Department of the Interior work toward early issuance of 10-year licensing schedules; (3) the Department of the Interior, in conjunction with appropriate states, prepare a regional programmatic impact statement when any coastal region is included on the 10-year licensing schedule; (4) an ad hoc committee be constituted by the Council on Environmental Quality to review regional programmatic impact statements; (5) in the absence of competing requests or if the character of the marine mineral deposits is not known and upon submission of an acceptable detailed exploration program the government issue an exploration license; (6) where competition exists for detailed exploration licenses, and where the Department of the Interior judges competing exploration plans to be technically sound, environmentally acceptable and the explorers capable of carrying them out, exploration licenses be given to (a) all parties, or (b) a cooperative exploration program be established by the Depart-

company without prior government approval or permit required; (2) the Department of the Interior prepare a regional programmatic impact statement in advance of any mineral licensing in an area; (3) it be government policy to license operations for mineral exploration and exploitation in the deep ocean; (4) marine mining licenses not be allocated on the basis of either bonus bids or royalty bids, but rather, royalty rates be established by the federal government based on the costs and benefits to the licensee; (5) based on the programmatic impact statement, the Department of the Interior, within a set time frame, prepare a license impact statement in advance of conversion to a production license; (6) conversion of the license from exploration to production be at the option of the licensee; (7) upon conversion of the license to the production phase, that government have access to all raw resource data and interpretive results; (8) to assist in setting royalty rates for new or renewed licenses, the Department of the Interior, at fixed intervals, establish ad hoc commissions to assess the adequacy of the royalties being charged mining companies; (9) at the time production operations begin, substantial percentages of the resource covered on the initial license be relinquished, in order to encourage extensive and early exploration and to provide maximum information; (10)

Outer Continental Shelf	Deep Ocean

Regulations and Leasing (cont'd)

ment of the Interior; (7) based on the programmatic statement for the region, the Department of the Interior prepare a license impact statement to be available at least three months prior to a public hearing on the proposed exploration program; (8) production licenses be given on the basis of work programs submitted by the companies with exploration licenses; (9) at fixed time intervals, substantial percentages of the land covered in the initial exploration license revert to the federal government; (10) after issuing a production license, government and the public have access to all raw technical data and interpretive results held by the licensee; (11) licenses not be allocated on the basis of bonus bids. Rather, royalty rates should be established by the federal government based on the costs and benefits to the licensee; in setting royalty rates for new or renewed licenses, the Department of the Interior should, at fixed intervals, establish ad hoc *commissions to assess the adequacy of royalties being charged mining companies; (12) the licensee should be liable for the consequences of his activities (see page 81); (13) except where prohibited by existing legislation, responsibility for resource management and regulation of offshore mining be concentrated in the Department of the Interior;*[7] *(14) the United States government support the establishment of an independent standard-setting organization, using Det norske Veritas as a*

except where prohibited by existing legislation, responsibility for the resource management and regulation of offshore mining be concentrated in the Department of the Interior; (11) the United States government support the establishment of an independent standard-setting organization, using Det norske Veritas as a model, with this organization in turn providing the technological backup for United States regulation of marine mining; and (12) all safety and environmental protection technology used in marine mining operations meet the best available commercial standard.

Outer Continental Shelf	Deep Ocean

Regulations and Leasing (cont'd)

Outer Continental Shelf	Deep Ocean
model; and (15) all safety and environmental protection technology used in marine mining operations meet the best available commercial standard.	

Education

Outer Continental Shelf	Deep Ocean
Conclusion. Sufficiently varied curricula at the university level are available for education in the supporting basics of marine mining, but that specialization is only obtained by student participation in marine minerals exploration or mining research projects. Basic technician training is widely available in two-year programs, but must be augmented by on-the-job experience at sea. Formal degree programs in ocean mining, per se, do not appear to be justified at this time, based on current activity.	Conclusion. Same.
Recommendations. *The Panel recommends that (1) government agencies responsible for management of marine mineral resources utilize existing non-government training facilities in meeting their needs for professional and technician-level personnel; (2) government-sponsored academic research and training in selected aspects of seafloor minerals exploration, marine mining and environmental considerations be continued and strengthened in cooperation with the academic and industrial sectors; and (3) an appropriate government agency initiate a study on existing and projected personnel requirements of*	Recommendations. *Same.*

Outer Continental Shelf | Deep Ocean

Education (cont'd)

the marine mining industry, including those associated needs of agencies and academia, to provide long-range educational guidance.

[1] U.S. President's Materials Policy Commission. 1952. Resources for Freedom: A Report to the President, Washington, D.C.: U.S. Government Printing Office.

[2] National Commission on Materials Policy. 1973. Material Needs and the Environment Today and Tomorrow. Final Report, Washington, D.C.: U.S. Government Printing Office; Committee on the Survey of Materials Science and Engineering. 1974. Materials and Man's Needs, Washington, D.C.: National Academy of Sciences; Proceedings of a Joint Meeting of the National Academy of Sciences-National Academy of Engineering. 1975. National Materials Policy, Washington, D.C.: National Academy of Sciences.

[3] Committee on Mineral Resources and the Environment, National Research Council. 1975. Mineral Resources and the Environment, Washington, D.C.: National Academy of Sciences.

[4] Det norske Veritas (Norway): is a major international ship classification society. It is independent and non profit.

[5] Amos, A.F., et al. 1972. Effects of Surface-Discharged Deep Sea Mining Effluent. Mar. Tech. Soc. Jour. 6 (4), pp. 40-46.

[6] Roels, O.A., et al. 1973. Environmental Impact of Deep-Sea Mining, NOAA Technical Report ERL 290 OD11, Boulder: Department of Commerce.

[7] U.S. Congress. Senate. Committee on Interior and Insular Affairs. 1975. Recent Developments in Deep Seabed Mining, 94th Cong., 1st session. Washington, D.C.: U.S. Government Printing Office.

TABLE OF CONTENTS

LIST OF TABLES AND FIGURES

MINING IN THE OUTER CONTINENTAL SHELF AND IN THE DEEP OCEAN

technological capabilities to work there; environmental considerations and regulatory and leasing requirements of outer continental shelf mining.

3. Deep-Ocean Mining:

   An assessment of geographical characteristics of the deep-ocean and of the technological capabilities of the industry to operate in it; the environmental implications of deep-ocean mining, as well as regulatory and leasing requirements that should apply.

4. Education and Manpower:

   A consideration of the educational issues germaine to both deep-ocean and outer continental shelf mining (to include technical training and stimulation of public awareness).

During the initial phase of the 18-month study, the Panel considered these issues. Comprehensive assessments were prepared by individual Panel members for careful review and analysis by the full Panel. Following extensive deliberations, a draft report was prepared.

In order to obtain additional viewpoints as to the validity of the problem areas, weaknesses and deficiencies in the assessments, and alternative options for dealing with the problems, the Panel convened a workshop, at which experts from government, industry, and academia participated. Subsequent to the workshop, the Panel integrated its findings into this final report.

MINING IN THE OUTER CONTINENTAL SHELF AND IN THE DEEP OCEAN

CHAPTER ONE

STUDY BACKGROUND

In August 1973, following a formal request by the Assistant Secretary for Mineral Resources of the Department of the Interior on 15 February 1973, the Marine Board of the National Academy of Engineering established a Panel on Operational Safety in Marine Mining for the purpose of assisting the Department of the Interior in formulating its policies concerning marine mining operations and management. The terms of reference developed by the Panel were as follows:

1. To identify the state of the art in marine mining exploration and recovery (including an assessment of technological capabilities, environmental considerations, and regulatory aspects of mining operations;

2. To examine the role of the U.S. government with regard to the development of the hard minerals of the outer continental shelf (OCS), with particular attention to the protection of the environment;

3. To identify engineering investigations needed to support the role of the government; and

4. To provide for wise use of the resources while minimizing the impact on the environment.

In examining these matters, the Panel found that limiting its view to operational safety--without consideration or assessment of other aspects of marine mining--would result in an imbalanced and incomplete study. By mutual agreement with the Department of the Interior, therefore, the Panel expanded the study to embrace the following:

1. <u>Importance and Potential of Hard Minerals from the Seafloor</u>:

   An assessment of resource availability and economic value of seafloor deposits; and a comparison of these resources with the availability and costs of terrestrial sources.

2. <u>Outer Continental Shelf Mining</u>:

   An assessment of geographical characteristics of the outer continental shelf and of industry's technological capabilities to work there; environmental considerations and regulatory and leasing requirements of outer continental shelf mining.

3. Deep-Ocean Mining:

   An assessment of geographical characteristics of the deep-ocean and of the technological capabilities of the industry to operate in it; the environmental implications of deep-ocean mining, as well as regulatory and leasing requirements that should apply.

4. Education and Manpower:

   A consideration of the educational issues germaine to both deep-ocean and outer continental shelf mining (to include technical training and stimulation of public awareness).

During the initial phase of the 18-month study, the Panel considered these issues. Comprehensive assessments were prepared by individual Panel members for careful review and analysis by the full Panel. Following extensive deliberations, a draft report was prepared.

In order to obtain additional viewpoints as to the validity of the problem areas, weaknesses and deficiencies in the assessments, and alternative options for dealing with the problems, the Panel convened a workshop, at which experts from government, industry, and academia participated. Subsequent to the workshop, the Panel integrated its findings into this final report.

CHAPTER TWO

MARINE MINERALS AND MINING: IMPORTANCE AND POTENTIAL

The importance of marine minerals to the economies of the United States and the rest of the world has increased dramatically since the oil embargo of 1973. The lessons in economics and politics administered by some of the oil producing nations have already resulted in price rises of another commodity in the world market--bauxite, the mineral used in making aluminum. Although the United States is more autonomous in nonfuel minerals than any country except the USSR and perhaps Canada, the present dependency on foreign sources, as shown in Table 2, is estimated to rise from $6 billion in 1971 to more than $50 billion by the year 2000.[8]

Recent computer models developed by Forrester[9] and Meadows, et al,[10] analyzing the impact of contemporary growth patterns on the world environment, use five factors basic to the human ecosystem: population, food production, industrialization, consumption of nonrenewable resources, and environmental pollution. With each increasing and interacting at a rapid rate, the models produced some interesting predictions. Increases in population require increases in food production, which requires increases in industrialization, leading to increased consumption of nonrenewable resources and increased pollution. Ignoring the other factors and allowing for a generous 250-year supply of natural resources, Forrester predicted that, well before the resources were exhausted, shortages would depress the world ecosystem and dynamic consequences would be felt in only 30-50 years, not 250 years in the future. Without necessarily endorsing, en toto, the specific results of these studies, the general conclusions of increasing shortages of various resources is concurred with. The major problem with the Forrester model, say many critics, is that it does not take proper account of technological innovation, engineering adaptations, substitute materials, etc.

If there is in fact a potential shortage of the mineral commodities required to meet future demands, what alternatives are open to the United States with regard to these deficits? There are several options, including reducing the rate of consumption through recycling or lowering the nation's growth patterns and living standards, finding alternate materials, and developing alter-

TABLE 2

IMPORTS SUPPLIED SIGNIFICANT PERCENTAGES OF TOTAL U.S. DEMAND IN 1973

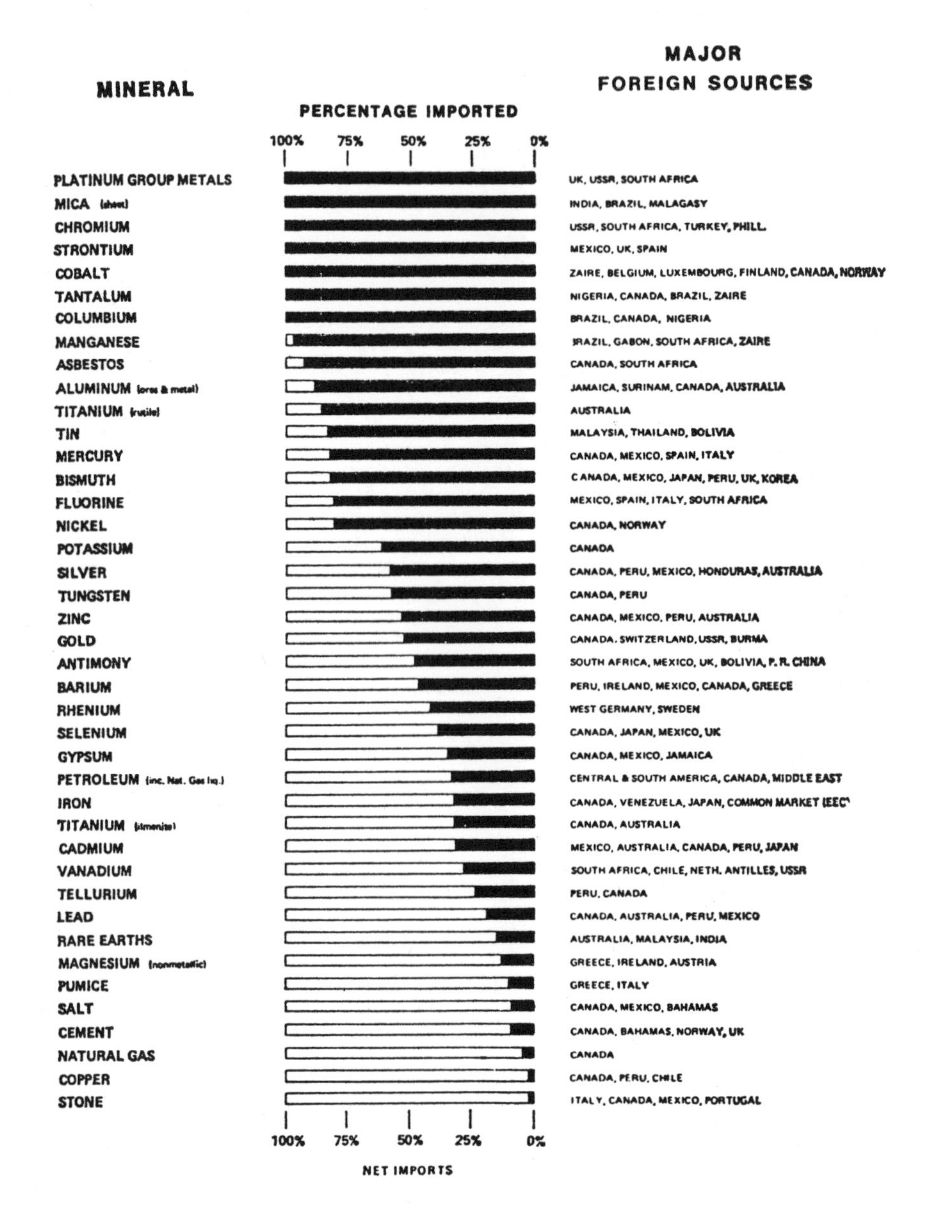

(Source: U.S. Bureau of Mines)

nate sources of minerals supply. These alternatives are not mutually exclusive, nor do they fall within the scope of this study, which examines only the specific case of marine minerals. That there is the potential for shortage is illustrated by Table 3.[11] This table presents the existing annual United States and world demand for 88 major mineral commodities, excluding oil and gas. To make comparison simpler, all numbers are reduced to Order of Magnitude dollars ($OM), where $OM8.862 represents $0.862 x 10^8.[12] The demands are extrapolated to give comparative estimated demands to the year 2000. These are then compared with the estimated total land resources of the commodities now in use by the United States and the world.

By the year 2000, the following commodity deficiencies are indicated for the United States:

- aluminum
- antimony
- asbestos
- barium
- bismuth
- cadmium
- cesium
- chromium
- cobalt
- copper
- diamond
- fluorine
- germanium
- gold
- graphite
- indium
- lead
- magnesium
- mercury
- mica
- nickel
- niobium (columbium)
- platinum
- quartz crystal
- sand and gravel
- silver
- sulfur
- tantalum
- tin
- tungsten
- uranium

World commodity deficiencies by the year 2000 are reckoned to include:

- aluminum
- abestos
- barium
- bismuth
- cadmium
- copper
- diamond
- fluorine
- germanium
- gold
- indium
- lead
- mercury
- sand and gravel
- sulfur
- tin
- tungsten
- uranium
- zinc

United States shortages of certain commodities are expected by the year 2000, not by the cutoff of foreign supplies, but rather by economic inaccessibility due to rising prices. The possibility of very large price increases arbitrarily imposed through the political actions of resource-rich countries can only be reduced if alternatives are available to the user. It is also important to consider the possibility that development of alternate sources might seriously damage the economy of a

TABLE 3. Total apparent world resources of marine minerals for dissolved, unconsolidated, and consolidated deposits, compared with estimated terrestrial resources, demands and adequacy of supply for U.S. and the world. (M. Cruickshank, *Technological and Environmental Considerations in the Exploration and Exploitation of Marine Minerals*)

Energy Resources	Symbol	Present Demand U.S.	Present Demand World	Cumulative Demand to 2000 AD U.S.	Cumulative Demand to 2000 AD World	Apparent Resources (Land) U.S.	Apparent Resources (Land) World	Adequacy (Land) U.S.	Adequacy (Land) World	Apparent Resources (Marine) Diss.	Apparent Resources (Marine) Unc.	Apparent Resources (Marine) Con.
Anthracite	C	8.862	10.173	10.208	11.549	11.580	13.297	+	+	NA		13.11
Bituminous coal and lignite	C	10.233	11.103	12.195	12.556	13.182	14.234	+	+	NA		13.87
Geothermal[1]	–	ND	ND	ND	ND	14.196	14.305	+	+	ND		14.11
Carbon	C	9.550	10.378	11.286	12.161	–	–	+	+	13.363		NA
Helium	He	8.253	8.268	10.267	10.311	11.281	11.281	+	+	ND		ND
Hydrogen	H	9.515	10.186	12.148	12.319	–	–	+	+	ND		ND
Peat	C	8.106	10.258	9.584	12.114	12.161	13.336	+	+	NA		NA
Shale oil	HC	–	8.900	11.600	11.822	13.909	15.951	+	+	NA		15.36
Thorium	Th	7.150	7.27	9.256	9.859	10.818	11.239	+	+	11.209	11.18	11.11
Uranium	U	8.509	8.924	11.450	11.833	11.309	11.384	–	–	12.578	14.28	11.16
Deuterium	–									(20.246)		ND
		11.067	11.347	12.86	13.223	14.223	15.931			13.423	11.16	15.38

1. Theobold *et al.*, 1972.

Note: High numbers (over 10^{14}) which would mask the general trends have been excluded from these totals.

Abbreviations:
INC Included in total.
NVA Present but no value assigned.
NA Not applicable.
ND No data.
NK Not known.

TABLE 3 (Continued)

Ferrous Minerals	Symbol	Present Demand		Cumulative Demand to 2000 AD		Apparent Resources (Land)		Adequacy (Land)		Apparent Resources (Marine)		
		U.S.	World	U.S.	World	U.S.	World	U.S.	World	Diss.	Unc.	Con.
Chromium	Cr	8.241	9.105	10.147	10.639	9.106	11.412	–	+	10.429	14.14	10.44
Cobalt	Co	8.262	8.817	10.124	10.369	9.603	10.890	–	+	11.227	14.22	11.35
Columbium	–	7.590	8.119	9.483	10.110	9.435	11.276	–	+	ND	12.26	11.90
Iron	Fe	11.129	11.659	11.669	12.342	12.277	13.174	+	+	10.236	NVA	12.32
Manganese	Mn	8.637	9.457	10.288	11.250	10.376	12.104	+	+	10.169	14.00	11.23
Molybdenum	Mo	8.904	9.224	10.596	11.159	11.102	11.175	+	+	12.497	14.86	10.13
Nickel	Ni	9.300	9.878	11.210	11.618	10.978	12.138	–	+	11.576	14.71	11.62
Rhenium	Re	6.400	7.100	8.367	8.840	9.232	12.609	+	+	ND		9.48
Silicon	Si	9.148	9.488	10.839	11.322	12.144	13.144	+	+	18.132	NVA	14.33
Tantalum	Ta	8.114	8.197	8.935	10.175	8.704	10.678	–	+	ND		11.18
Tungsten	W	8.427	9.197	10.450	11.141	10.165	10.879	–	–	ND	12.71	10.22
Vanadium	V	8.196	8.578	10.229	10.537	11.123	11.425	+	+	11.156	15.87	12.15
		10.203	10.911	12.116	12.510	12.460	12.511			14.138	16.16	14.34

TABLE 3 (Continued)

Nonferrous Minerals	Symbol	Present Demand		Cumulative Demand to 2000 AD		Apparent Resources (Land)		Adequacy (Land)		Apparent Resources (Marine)		
		U.S.	World	U.S.	World	U.S.	World	U.S.	World	Diss.	Unc.	Con.
Aluminum	Al	10.198	10.525	12.295	12.832	10.678	11.678	–	–	13.390		11.77
Antimony	Sb	8.193	8.632	10.122	10.339	9.266	10.393	–	+	11.280		8.76
Arsenic	As	7.380	7.840	9.182	9.412	9.304	9.672	+	+	11.490		9.13
Beryllium	Be	8.407	8.593	10.307	10.492	11.134	12.190	+	+	ND	NVA	12.10
Bismuth	Bi	7.920	8.304	9.495	10.158	9.154	9.870	–	–	11.245	12.45	9.66
Cadmium	Cd	8.353	8.830	10.316	10.832	9.554	10.381	–	–	10.806	NVA	9.40
Cesium	Cs	–	–	6.400	6.700	0	9.132	–	+	10.919		11.23
Copper	Cu	10.130	10.774	12.168	12.697	11.684	12.287	–	–	11.129	13.62	11.16
Gallium	Ga	6.400	7.110	8.236	8.705	10.324	11.300	+	+	13.834	NVA	13.67
Germanium	Ge	7.200	7.920	9.131	9.630	8.123	9.240	–	–	ND	NVA	12.13
Gold	Au	9.259	10.116	11.214	11.773	11.117	11.470	–	–	12.246	13.91	10.20
Hafnium	Hf	7.310	7.740	9.204	9.493	11.213	11.527	+	+	ND		12.35
Indium	In	7.150	7.530	8.739	9.273	8.318	9.196	–	–	ND		10.32
Lead	Pb	9.243	9.931	11.133	11.445	10.837	11.278	–	–	11.413	12.47	10.17
Magnesium	Mg	9.144	9.511	11.346	11.589	11.106	13.182	–	+	17.141	NVA	13.61
Mercury	Hg	8.332	9.138	10.362	11.153	9.500	10.803	–	–	10.634		10.29
Platinum-group metals	–	9.202	9.643	11.141	11.496	9.583	11.824	–	+	ND	10.17	10.96
Radium	Ra	–	–	–	–	10.117	11.117	+	+	ND		11.13
Rare-earth elements	–	8.204	8.354	10.150	10.308	11.171	11.306	+	+	11.138	13.19	12.25
Rubidium	Rb	–	–	6.100	6.200	6.900	8.288	+	+	13.166		6.37
Scandium	Sc	–	–	7.230	7.260	9.351	10.351	+	+	ND	NVA	13.81
Selenium	Se	7.550	8.134	9.245	9.659	9.255	9.964	+	+	12.628		9.38
Silver	Ag	9.193	9.728	11.132	11.496	11.105	11.231	–	+	12.287	9.24	10.26
Tellurium	Te	7.130	7.238	8.594	9.119	9.205	9.739	∓	+	ND	13.34	7.89
Thallium	Tl	–	6.100	7.190	7.84	7.199	8.110	+	+	ND		11.62
Tin	Sn	9.196	9.819	11.322	11.653	9.153	11.303	–	–	11.136	12.69	9.96
Titanium	Ti	9.414	10.127	11.385	12.115	11.667	12.388	+	+	10.404	12.74	13.26
Yttrium	Y	7.340	7.450	9.291	9.407	10.146	11.120	+	+	12.514	12.20	13.19
Zinc	Zn	9.380	10.146	10.241	12.109	10.810	11.335	+	–	11.413	14.91	10.72
Zirconium	Zr	8.342	8.	9.447	10.604	11.375	12.104	+	+	ND	11.19	–
		10.553	11.210	12.680	13.214	12.290	13.326			14.159	14.21	14.18

TABLE 3 (Continued)

Metallic Minerals	Symbol	Present Demand U.S.	Present Demand World	Cumulative Demand to 2000 AD U.S.	Cumulative Demand to 2000 AD World	Apparent Resources (Land) U.S.	Apparent Resources (Land) World	Adequacy (Land) U.S.	Adequacy (Land) World	Apparent Resources (Marine) Diss.	Apparent Resources (Marine) Unc.	Apparent Resources (Marine) Con.
Argon	Ar	8.269	8.681	10.206	10.495	<00	<00	+	+	ND		NA
Asbestos	–	8.704	9.302	10.374	11.200	9.172	9.862	–	–	0		9.32
Barium	Ba	8.204	8.574	10.102	10.398	9.887	10.222	–	–	11.121	11.10	10.43
Boron	B	8.411	9.113	10.292	10.866	11.170	11.340	+	+	14.333	12.26	9.58
Bromine	Br	8.845	9.113	10.476	10.713	<00	<00	+	+	15.576		9.70
Calcium	Ca	9.364	10.111	11.238	11.749	<00	<00	+	+	15.235	15.42	11.58
Chlorine	Cl	9.607	10.136	11.483	12.116	12.360	<00	+	+	17.209		10.59
Clays	–	9.240	10.151	11.195	12.103	11.848	12.410	+	+	ND		–
Corundum and emery	–	6.400	7.9	8.294	9.150	8.720	10.100	+	+	0		–
Diamond	–	8.444	9.398	10.442	11.373	0	10.428	–	–	0		ND
Diatomite	–	8.275	9.103	10.207	11.102	11.348	12.116	+	+	ND	13.64	–
Feldspar	–	7.820	8.263	9.627	10.202	10.620	11.124	+	+	ND		10.46
Fluorine	–	8.646	9.180	10.527	11.166	9.540	10.388	–	–	12.199		11.33
Garnet	–	7.200	7.250	9.183	9.235	9.735	10.368	+	+	0	10.76	–
Gem stones	–	9.510	10.114	ND	11.342	ND	ND	+	+	0		–
Graphite (natural)	–	7.300	8.265	9.173	10.196	8.500	10.520	–	+	0		10.19
Gypsum	–	8.574	9.204	10.330	11.118	11.734	12.367	+	+	0		12.14
Iodine	I	7.530	8.149	9.402	10.113	11.189	11.945	+	+	11.181		9.30
Kyanite and related minerals	–	7.940	8.260	9.858	10.239	10.250	10.886	+	+	0	12.12	10.33
Lithium	Li	7.490	7.820	9.391	9.637	10.767	11.111	+	+	13.282		11.23
Mica	–	7.760	8.193	9.338	10.485	Low	NK	–	NK	0		–
Nitrogen	N	9.702	10.245	11.398	12.180	12.345	12.368	+	+	12.176		10.11
Oxygen	O	9.139	9.454	10.751	11.290	<00	<00	+	+	ND		NA
Perlite	–	7.410	7.890	9.280	9.685	10.790	11.296	+	+	0		–
Phosphorus	P	9.156	9.516	11.120	11.504	12.306	12.981	+	+	11.482	11.52	11.22
Potassium	K	9.119	9.478	11.104	11.471	11.127	13.366	+	+	15.140	11.20	12.36
Pumice	–	7.61	8.237	9.471	10.200	10.158	11.158	+	+	–		–
Quartz crystal	–	6.200	6.400	7.640	8.778	0	NK	–	NK	0		–
Sand and gravel	–	10.102	10.745	11.767	12.555	11.666	12.333	–	–	0	14.31	–
Sodium	Na	9.308	10.101	11.223	11.799	12.240	<00	+	+	16.301	NVA	12.21
Stone	–	10.122	10.845	11.915	12.648	<00	<00	+	+	0		–
Strontium	Sr	6.300	6.600	8.252	8.488	8.554	9.245	+	+	13.977		10.91
Sulfur	S	9.382	10.144	1.253	12.104	11.128	12.104	–		15.504	NVA	10.81
Talc, soapstone, & pyrophyllite	–	7.620	8.318	9.407	10.274	9.623	10.350	+	+	0		–
Vermiculite	–	7.550	7.820	9.374	9.805	10.588	11.588	+	+	0		–
		10.585	11.302	12.416	13.229	12.936	13.630			15.348	15.46	12.88

nation heavily reliant on raw material exports to maintain a balance of trade.

Means of accommodating the increasing demands are being implemented, and include methods such as:

1. exploration and discovery of new ore bodies
2. development of new mines
3. increasing production of existing mines
4. working and reopening known deposits of lower grade ore and tailings
5. recycling of materials
6. development of more efficient processes to convert ores and metals into useful products.

The development of marine resources is important to the maintenance of the international economic and political balance and to support the standard of living in the United States. While it is probably not feasible or desirable for the United States to become self-sufficient for the basic mineral commodities, the Panel considers it prudent to develop adequate alternate sources of supply from the sea.

Estimates of apparent marine mineral resources have been developed by M. Cruickshank for dissolved, unconsolidated, and consolidated deposits (Table 3). With the exception of asbestos, graphite, and quartz crystals, where data are available and deficiencies have been predicted, alternative marine sources for the minerals exist and may exceed existing land resources. While few of these reserves have been positively identified at the present time, certain specific commodities have been found along the outer continental shelf and on the deep seabed. As marine mining and extractive technology are developed, it is believed that these apparent resources will become viable mineral sources.

EVALUATION OF POTENTIAL

Despite considerable recent interest in the mineral potential of the seafloor, the number of marine mineral deposits now being utilized, even from under the relatively accessible waters of the world's continental shelves, is small. This situation is particularly true in the case of the United States continental margins where only a limited number of marine mining activities have taken place. These include the working of hard rock barite deposits and gold placer deposits off Alaska's shore, the Grand Island Frasch sulfur deposits in the Gulf of Mexico, and the deposits of sand, gravel, and calcium carbonate (shell, coral, and aragonite along the East and Gulf Coasts). Marine mining activities can be expected to increase, as the demand and cost of land-based natural resources increase. Table 4 provides a classification of dissolved, unconsolidated, and consolidated resources that are known to be in the

TABLE 4. Classification of marine mineral resources. (M. Cruickshank and R. Marsden, Marine Mining: Section 20, SME Mining Engineering Handbook, Volume 2; I.A. Given and A.B. Cummins, eds.), AIME, New York, 1973.

Marine Mineral Deposits

	Unconsolidated			
Dissolved	Continental Shelf, 0-200 M	Continental Slope, 200-3,500 M	Deep Sea, 3,500-6,000 M	Consolidated
Seawater:	*Nonmetallics:*	*Authigenics:*	*Authigenics:*	*Disseminated, massive, vein, tabular, or stratified deposits of:*
Fresh water	Sand and gravel	Phosphorite	Ferromanganese nodules and assoc.	Coal
Metals and salts of:	Lime sands and shells	Ferromanganese oxides and assoc. minerals	Cobalt	Ironstone
Magnesium	Silica sand	Metalliferous mud with:	Nickel	Limestone
Sodium	Semiprecious stones	Zinc	Copper	Sulfur
Calcium	Industrial sands	Copper	*Sediments:*	Tin
Bromine	Phosphorite	Lead	Red clays	Gold
Potassium	Aragonite	Silver	Calcareous ooze	Metallic sulfides
Sulphur	Glauconite		Siliceous ooze	Metallic salts
Strontium	*Heavy Minerals:*			Hydrocarbons
Boron	Magnetite			
Uranium	Hmenite			
Other elements	Rutile			
Metalliferous Brines:	Monazite			
Concentrations of:	Chromite			
Zinc	Zircon			
Copper	Cassiterite			
Lead	*Rare & Precious Minerals:*			
Silver	Diamonds			
	Platinum			
	Gold			
	Native copper			

ocean. Dissolved deposits are contained in the seawater and may be considered to include dissolved minerals, such as bromine and magnesium, and biogeochemical concentrations in certain plants and animals, such as iodine in seaweeds or phosphatic compounds in fish skeletons.

Unconsolidated deposits may be defined as those surface or near surface deposits on the seabed amenable to dredging. These occur at all depths of the ocean and include placer deposits of gold, heavy minerals, and construction materials, such as sand, gravel and lime shells, generally found in relatively shallow water. Surficial deposits of phosphorite and ferromanganese oxides are found as loose nodules on the seafloor in deeper waters. Other deposits of marine origin include vast concentrations of siliceous and calcareous oozes and unusually high concentrations of metalliferous muds associated with active regions of seafloor spreading.

Consolidated deposits occurring as hard rock on the continental shelf may be as prolific and diverse as the familiar mineral deposits mined on land. In the deep seabed such deposits may include encrustations or indurations of metalliferous ferromanganese oxides and possibly other concentrations associated with geoactivity on the seafloor. Detailed knowledge of marine deposits is very small by comparison with the resources that may be projected by statistical inference.

Exploring for minerals in the seas is very difficult and complex. Only a small percentage of the continental margins and deep-ocean basins have been surveyed for hard minerals. Before the full potential of these minerals can be realized, the technological and engineering capabilities to locate and assess them must be improved and applied. Based upon present geological understanding of the nature of the continental margins and deep seabeds, however, substantial deposits remain to be discovered.

Minerals on the United States continental shelves that possess the potential for early economic development are the surficial deposits of sand, gravel, and calcium carbonate, placer deposits of titanium and gold, and marine phosphorite deposits.

Sand and gravel deposits, which provide excellent potential sources of low cost building aggregate and road materials lie seaward of many United States cities. Sand for beach replenishment in recreation areas is abundant in many offshore regions and can serve as a substitute for dwindling land sources.

Calcium carbonate is being recovered frcm the seabed on the continental shelves in the form of shells, shell sands and muds, aragonite, and coral.

Placer deposits are concentrations of minerals produced by the action of moving waters in rivers, waves on beaches, or tidal currents. Many submerged placers on the ocean floor are extensions of on-land placer deposits. Minerals which may be concentrated in placer deposits include gold, platinum, and ores of tin, iron, titanium, chromium, and zirconium. A promising area for seabed gold placer depcsits is in Norton Sound, Alaska. Gold is also known to exist in submerged deposits off Oregon and California.

Submarine phosphorite occurs in many scattered localities off the coasts of continents. The favored environment is in areas of nutrient-rich upwelling waters where detrital sedimentation is low. The principal offshore phosphorite nodule deposits of the United States continental margin are in the California borderland region off southern California. These deposits may prove valuable to the nation's economy by augmenting the known reserves and by fulfilling the needs of local markets. Recent increases in the price of imported phosphates may result in increased interest in the exploitation of these deposits. Recognizing the possibility that leasing applications for both oil and gas operations, and, in this case, phosphorite, might occur in the same location, special attention would need to be given to leasing and regulatory questions.

It is believed that large deposits of copper, nickel and other metals may be present in bedrock beneath continental shelves and could conceivably be mined in the future. Little is known, however, about the potential of these buried consolidated rock deposits. In addition, the nature and potential of the deep ocean metalliferous muds are in an early stage of understanding, and the commercial possibilities of these deposits, except for some in the Red Sea, have not been studied.

Beyond the continental margins, in the deep-ocean basins, manganese nodules and associated crusts and pavements contain the deep ocean minerals most likely to be exploited in the near future. Bearing fine-grained metal oxides, these are distributed widely over the floor of the world's oceans. They vary widely in their composition, as well as in their physical and chemical properties. Considerable commercial activity exists in developing these resources for their major component metals, chiefly nickel, copper, cobalt and manganese.

To date, only a very small percentage of the deep seabed has been surveyed extensively but enough has been learned about the extent and location of these surficial deposits to permit the first stages of commercial development to begin. On the basis of known concentrations, average compositions, and operating conditions, nodules from the northeast Pacific have attracted the greatest attention. These nodules are generally located in water depths of

2600 to 5500 meters (12,000 to 18,000 ft) and lie on deep ocean siliceous and red clay sediments.

Access to these resources is important to the United States and should be made available on fair and equitable terms. The conditions under which deep seabed resources will be available to nations is a major topic under consideration at the United Nations Law of the Sea conferences. The principal technology for developing nodule resources resides with several United States firms and this leadership is evidenced by the present activities.

In order to operate successfully on the seafloor for the purpose of collecting or moving large amounts of materials, it is important to have a knowledge of the nature and distribution of the materials being collected, the environment in which the materials exist, and the conditions under which such operations will take place. The means must, therefore, be available to locate and delineate the extent of a deposit and to determine the important properties of the associated marine sediments that will bear on the design of the machines to operate therein. In addition to the properties of the seafloor and associated terrain, the environmental loads imposed by the water column will have a major influence on the design of the engineering structures needed to mine the deposits. Most of the nodules occur at the sediment-water interface and, accordingly, there is no need to penetrate the substrate during dredging operations.

In evaluating marine mining potential, two basic constraints must be considered:

1. <u>Resource Assessment</u>

Technical capabilities to determine the important parameters associated with continental margin resource assessment have not progressed appreciably in the last five to ten years. In the deep ocean, however, several new tools have been developed and these have permitted the discovery of many potential ore bodies. A most important need is to develop new and improved instruments and equipment to conduct rapid and effective surveys over wide areas of the continental margins that will lead to the initial discovery of potential ore bodies in these areas. However, before new tools can be developed, a better understanding of marine placer depositional mechanisms must be obtained in order to define the requirements for equipment performance.

To provide for safe and cost-effective reconnaissance surveys of coastal waters, remote sensing systems are needed that can provide information on sediment dispersal patterns, currents, and potential mineral sites. Improved coring tools and techniques for mineral sampling and evaluation are also needed. The available devices for coring and spatial sampling of potential ore bodies are expensive, slow in collecting samples, and require specialized ship facilities. Rotary coring systems in use on survey ships has not yet proven to be reliable or effective in all instances. Short coring tools, such as the gravity corer and the piston corer, are of little value because they are limited to shallow penetration, and, therefore, deeper resources remain undiscovered. The rotary tool has been more successful on hard sand and gravel bottoms, but it causes considerable disturbance of the sample. The need exists for new approaches that include further developments of vibratory corers, pressure-jetting hydraulic systems, or corers based on the principle of electroosmosis.

In the deep ocean, one of the most important needs is the development of fully automatic systems to produce reconnaissance-scale topographic, geophysical and geological maps. Existing tools, including deep ocean television systems and cameras, wide beam bathymetric systems, and free fall samplers, have been used to locate potential ore bodies; however, the costs have been very high. New tools and techniques are needed to provide an order of magnitude increase in capabilities.

Bathymetric measurements, taken from the surface by small ships equipped with wide beam sonar transducers, cannot provide realistic representations of the deep ocean seafloor in which the manganese nodules of paramount interest occur. The development of a commercial version of the Navy's narrow beam bathymetric system, in association with an acoustic seafloor navigation array, is needed to permit the preparation of micro-bathymetric charts from a small survey ship operating at 8-15 knots. Data will need to be automatically processed on shipboard or ashore to produce contour maps for direct exploration on a real time basis. Micro-topographical maps, necessary for detailed resource assessment and mine development, will have to be improved. A system with such capability might be developed from the Scripps Institution's "deep-tow" system. It uses a narrow-beam echo sounder for topographic measurements, side-scan sonar, with several scales of resolution, for determining such bottom features as location and shape of rock outcrops and small escarpments, a 3-1/2 kHz echo sounder to delineate the details of shallow structures; stereo photographic equipment and a proton magnetometer. In addition, an acoustic transponder navigation system enables accurate positioning.

In order to advance the state-of-the-art of deep ocean resource assessment, there is a need for (1) the capability to conduct <u>in-situ</u> metal analysis and evaluation, (2) improved side-scan and sector-scan sonar equipment, (3) improved television systems to allow for relatively high speed towing (3-5 knots), (4) automatic television signal scanning apparatus to provide a nodule census, and (5) improved sample collection devices, operable from the surface and capable of collecting a large number of samples at known discrete locations to permit statistical determination of the percentage of various metals in the nodules.

2. Environmental Baselines

Depending on the type of deposit and its geographic location, each mining operation will have a different set of environmental parameters that must be determined to define the interactions between the environment and the mining system. The offshore petroleum industry has long recognized the importance of the acquisition of natural environmental information for use in the safe and effective design of many types of drilling structures and other equipment. Environmental data collected mostly by private firms are proprietary and generally only available at considerable cost. Much of the available information is limited to areas in the Gulf of Mexico where the major oil activities are concentrated.

Oceanographic and meteorological data in areas where marine mining is likely to occur needs to be acquired for the determination of environmental loads. These data will enable engineers to design ocean mining structures that are safe for personnel and the environment as well as more efficient and cost-effective.[13] Government regulation of marine mining will require full consideration of the physical environment and environmental tradeoffs.

PROBABLE AREAS OF EARLY OUTER CONTINENTAL SHELF MINING

Based on geological and geophysical surveys performed by government and academic organizations in the past few years and the limited amount of industry information made available, the Panel considers that there are several sites with near-term (within the next two decades) commercial mining potential. Among these are:

1. Gulf of Maine -- potential for lode deposits, chiefly sulfides in shallow waters, and some potential for sand and heavy minerals.

2. Massachusetts Coast -- parts of Cape Cod Bay and Buzzards Bay possess good potential for sand, rare earth heavy minerals, and possibly coal.

3. New Jersey-New York Bight -- known sand deposits.

4. Southeast Atlantic Coast -- known beach resource of heavy mineral sands, but the sand potential of the seaward outer continental shelf lands is incompletely known.

5. Gulf of Mexico -- potential for hard minerals on the outer continental shelf appears to be limited as a

whole, although the United States Geological Survey has identified abundant black sands (including titanium-bearing minerals) off the Texas coast; oyster shells may prove to be a resource on the outer continental shelf in the future. There is a possibility for finding economically attractive deposits of finely divided metal sulfides that were formed in place on the continental slope.

6. Southwest Pacific Coast -- known deposits of sand, gravel, and phosphorite.

7. Northwest Pacific Coast — outer continental shelf off northern California and Oregon is known to have modest placer deposits of gold and other heavy metals.

8. Great Lakes — although not outer continental shelf lands, the portion of the lake beds within the United States are known to have manganese and copper ore deposits.

9. Bering Sea -- outer continental shelf has the most promising potential for mining hard minerals of all United States outer continental shelf waters. Placer deposits of this potential include gold, platinum, cassiterite (tin), scheelite (tungsten), rare earths, ilmenite (titanium), and others. Lode deposits are likely to include barite and copper, lead and zinc (as sulfides), and molybdenum, while deposits of chemical precipitates of uranium-bearing minerals are probable in some anoxic sites. Government, industry, and academic groups have been conducting hard mineral surveys in this area for more than ten years.

10. Arctic Shelf — largely of unknown potential, but drainage from metal-bearing provenance rocks probably washes some noble metals into outer continental shelf high-energy sand sites.

11. Insular States and Territories -- although few mineral surveys have been made in the outer continental shelf, or its equivalent waters off American Samoa, Puerto Rico, Hawaii, or the Trust Territories, the potential for volcanic and basalt-related minerals and manganese crusts appears likely.

Thus, the outer continental shelf of the mainland United States north of Virginia on the East Coast, the West Coast, and the Gulf of Mexico, and the shelf in the central and northern Bering Sea, offer the prime exploratory regions for mineral mining. These areas should receive attention. Although presently considered to be marginal, secondary sites for hard minerals include the outer continental shelf lands off the Carolinas and the southeast Atlantic Coast, and off Texas. Much exploration and inventory remains to be done. The several worldwide potential sites reviewed by Moore in 1972[14] indicate that the United States stands in a very favorable position with regard to the probable wealth of hard minerals on its outer continental shelf.

8 Bureau of Land Management. 1973. Mining and Minerals Policy, Washington, D.C.: U.S. Department of the Interior.

9 Forrester, Jay W. 1971. World Dynamics, Boston: Wright-Allen Press.

10 Meadows, D.H., et al. 1972. The Limits to Growth, New York: New American Library, Inc.

11 Cruickshank, M.J. 1975. Technological and Environmental Considerations in the Exploration and Exploitation of Marine Minerals, Ph.D. dissertation, Madison: The University of Wisconsin.

12 The amounts on this table range from $200,000 ($0.2 \times 10^6$) to $24.6 million trillion ($24.6 \times 10^{18}$). For comparison, all numbers are reduced to exponential fractions and expressed as Order of Magnitude Dollars ranging from $0M6.200 to $0M20.246, using this notation.

13 Marine Board, National Research Council. 1975. Information and Data Exchange for Ocean Engineers: An Approach to Improvement, Washington, D.C.: National Academy of Sciences.

14 Moore, J.R. 1972. Exploitation of Ocean Mineral Resources - Perspectives and Predictions. Proceedings of the Royal Society of Edinburgh, Vol. 72, pp. 193-206.

CHAPTER THREE

OUTER CONTINENTAL SHELF MINING

The continental shelf is usually defined as the gently sloping, shallow water platform that extends from the coasts to the shelf break, after which the continental slope descends relatively steeply to the deep ocean floor. The worldwide average width of a shelf is 71 kilometers (44 miles), and the average termination depth is 140 meters (450 ft),[15] although the accepted legal definition of the termination depth is 200 meters (650 ft). The continental slope marks the submerged structural edge of the continents and overlies the transition area from thick continental crust to thin oceanic crusts.[16] Shelf and slope together are often referred to as the continental terrace. Much less is known about the sediments or bathymetry of the slope than of the shelf. Recent research interest, coupled with such new techniques as acoustic profiling, is increasing our knowledge of these areas.

The inner continental shelf has been defined as extending from shore to the 5 kilometer (3 mile) limit of the territorial seas.

Placer-type deposits, such as sand and gravel, may be found most typically on the shelf in water depths of the order of 92 meters (300 ft) or less, whereas the phosphorite deposits generally exist at water depths greater than 92 meters (300 ft), and in certain instances seaward of the continental shelf as deep as 457-762 meters (1,500-2,500 ft).

It is anticipated that in-situ deposits of hard minerals in the bedrock of the continental shelf occur with the same frequency that they appear under similar geologic conditions on land. Until now, few deposits have been located on the continental shelves of the United States, probably because surveys to define the deposits have not been made. In other parts of the world, deposits of coal, iron ore, and tin have been worked underground in relatively shallow water, chiefly as seaward extensions of known coastal deposits. In the United States, economic sulfur deposits have been located in the outer continental shelf, using exploration methods applicable to the search for oil.

The mineral deposits of interest differ from the surrounding seafloor materials, and their presence may be detected by measuring distinctive physical properties such as density, seismic velocity, magnetism, electrical and heat conductivity, and induced polarization or chemical properties. Useful methods of measurement need to have adequate resolution and penetration to detect the size of the deposit and the depth at which it lies below the bottom. In the usual exploratory techniques, a two-dimensional profile of the quantity is measured along a track across a region where the mineral deposits are considered to exist. The measurements are conventionally presented on a contour map which joins points of equal value, thereby outlining areas of high and low values.

There are several exploratory methods used on the continental shelf. Some provide direct information about the shape and composition of a deposit. These include vertical and horizontal echo sounders to produce acoustic images; underwater still and television cameras; corers, dredges, and drills. Others provide indirect quantitative measurements of the physical properties of consolidated and unconsolidated materials. These include seismic reflection; seismic refraction; magnetic; gravity; electric; heat flow, and radioactive. Still others indicate the presence of deposits by analyzing the constituents in unconsolidated and consolidated samples of the seafloor sediment. These include flame and arc emission spectroscopy; atomic absorption spectroscopy; x-ray fluorescence; electro-chemical methods and calorimetry.

Surveying the seafloor for mineral deposits requires accurate navigation techniques if worthwhile delineation of deposits and maps are to be produced. Methods for accurate positioning are available in areas which are within 320 kilometers (220 miles) of land stations. These areas are generally those associated with continental margin deposits.

TECHNOLOGICAL ASSESSMENT

The objectives of this section are:

1. to predict how mining may develop on the United States continental shelf and slope over the next two decades;

2. to describe the technology which may be used, and

3. to define the manner in which the various mining techniques may impinge on the marine environment.

Outer continental shelf mineral deposits can be mined in more than one way. Likewise, a given mining technique may have application to more than one type of mineral deposit. This relationship is shown in Table 5.

In order to anticipate the types of probable outer continental shelf mining operations that will de-

TABLE 5. Mining Techniques versus outer continental shelf deposit types.

MINING TECHNIQUES	CONTINENTAL SHELF		CONTINENTAL SLOPE	
	UNCONSOLIDATED DEPOSITS	CONSOLIDATED DEPOSITS	UNCONSOLIDATED DEPOSITS	CONSOLIDATED DEPOSITS
TRAILING SUCTION HOPPER DREDGE	SAND/GRAVEL SHELLS			
SUCTION DREDGE, ANCHORED	HEAVY MINERALS		PHOSPHORITE	
CUTTERHEAD PIPE-LINE DREDGE	SAND/GRAVEL SHELLS			
BUCKET LADDER DREDGE	HEAVY MINERALS			
DRAG DREDGE			PHOSPHORITE	
CONTINUOUS LINE BUCKET			PHOSPHORITE	
UNDERGROUND MINING		VARIOUS		VARIOUS
SOLUTION MINING		VARIOUS		VARIOUS

velop, six hypothetical, but possibly typical mining operations for various mineral deposits are described in the following pages. In addition to a description of the mining equipment and its production cycle, three related matters are considered: the preproduction time frame, the likely deposit size, and the environmental impact potential.

The six hypothetical mining operations are described in the following sections:

I. Sand and Gravel (Case I)
II. Shells (Case II)
III. Phosphorite (Case III)
IV. Heavy Minerals (Case IV)
V. Underground Mining (Case V)
VI. Solution Mining (Case VI)

Sand and Gravel (Case I)

Sand and gravel, utilized primarily in construction work, but also for the restoration of storm-damaged beaches and for waterfront fill, appear to offer the main potential for continental shelf mining. At present, however, except for a few relatively small operations in bays, tidal rivers, estuaries, and large lakes, most United States sand and gravel aggregate comes from land-based operations. The annual nationwide production is approximately 920 million metric tons (1 billion tons). By the year 2000, projections show the probable annual demand to be 3 to 4 billion metric tons (3 to 4 billion tons). While inland resources are virtually limitless, there is an imbalance between the distribution of the resources and the markets. Transportation plays an important part in the economics of production; truck transportation for as few as 40 kilometers (25 miles) can double the cost to the consumer. The problems of land production are not limited solely to resource availability and economics. Urban sprawl, zoning laws, and environmental constraints have limited the use of many sand and gravel deposits.

Where metropolitan areas, navigable waters, and favorable marine geology occur in juxtaposition, the continental shelf offers potential for adding to the nation's sand and gravel resource base. This has occurred in Europe, where eight nations annually mine more than 36 million metric tons (40 million tons) of sand and gravel from the North and Baltic seas. The United Kingdom supplies 16 percent of its need for construction aggregate from offshore. Similarly, Japan annually mines over 55 million metric tons (60 million tons) of sand, or 19 percent of its total needs.

Favorable areas off the United States coast are shown in Figures 1-3. The specific market areas of interest in-

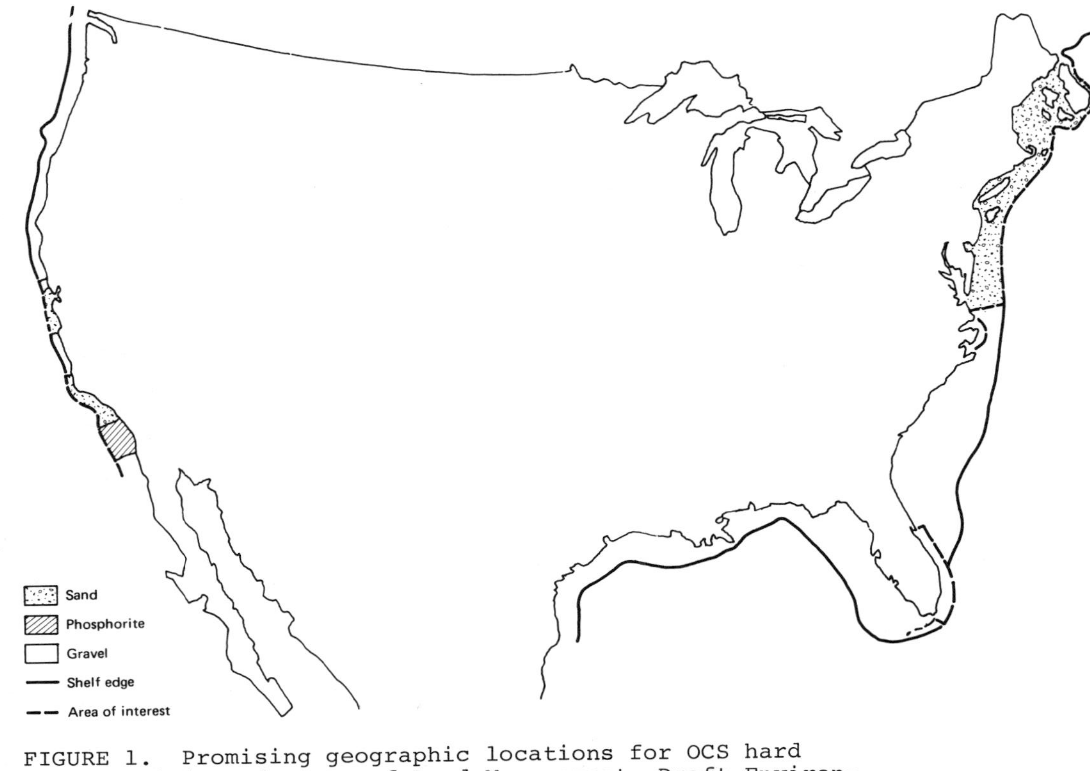

FIGURE 1. Promising geographic locations for OCS hard mineral mining. (Bureau of Land Management, Draft Environmental Statement - Proposed Outer Continental Shelf Hard Mineral Mining Operating and Leasing Regulations)

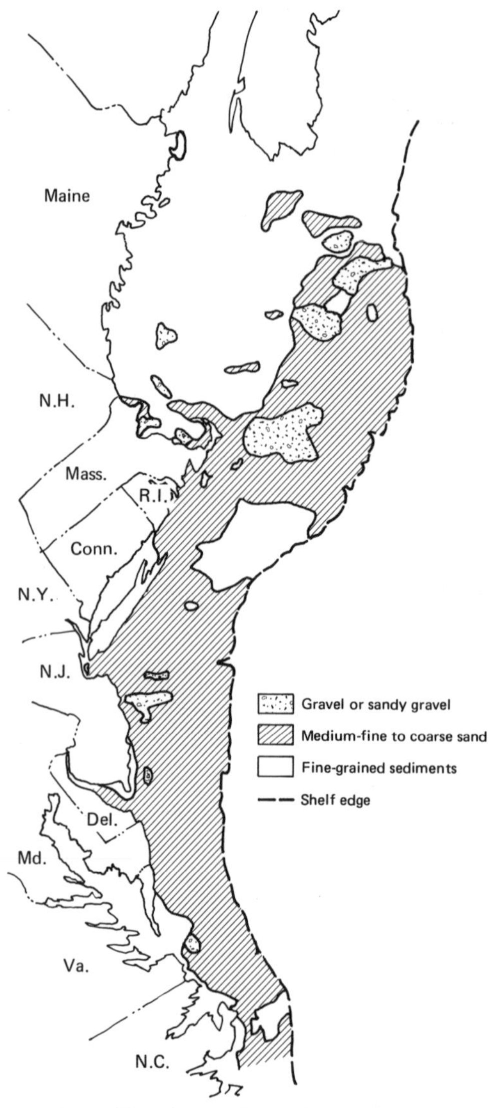

FIGURE 2. Distribution of sediments on the continental shelf, North Carolina to Maine. (Bureau of Land Management, Draft Environmental Statement - Proposed Outer Continental Shelf Hard Mineral Mining Operating and Leasing Regulations). Adapted from Schlee (1968), Schlee and Pratt (1970), and Milliman (1970).

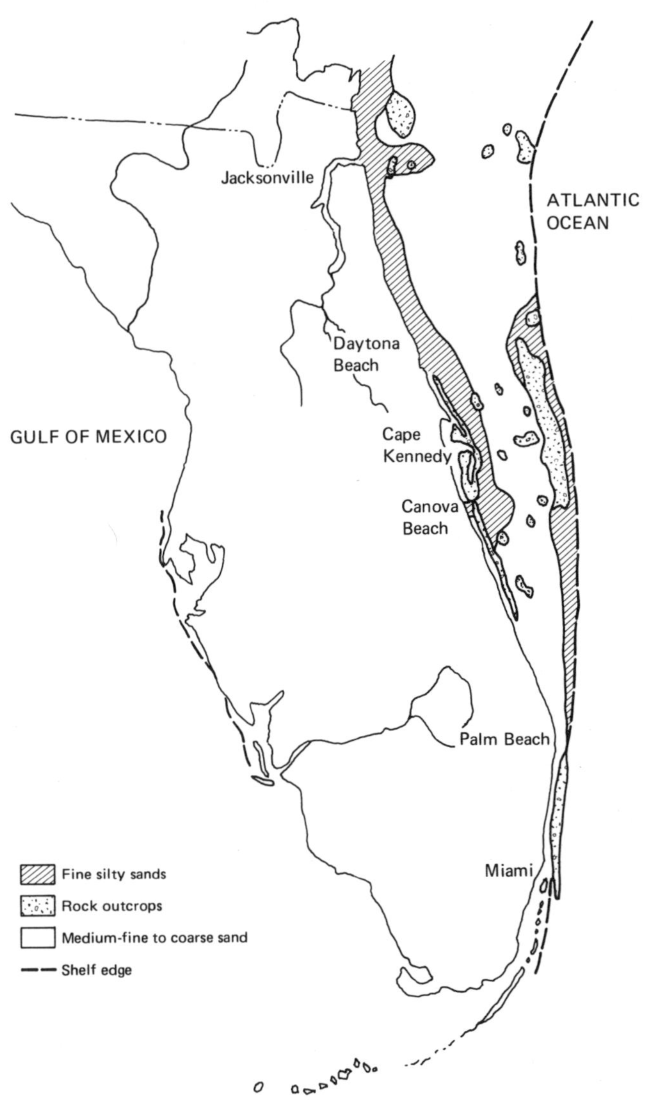

FIGURE 3. Distribution of sediments, Florida Atlantic continental shelf. (Bureau of Land Management, Draft Environmental Statement - Proposed Outer Continental Shelf Hard Mineral Mining Operating and Leasing Regulations). Compiled from Milliman (1972), and Uchupi, (1968).

clude Boston, New York, Washington, D.C., Norfolk, Southeastern Florida, Los Angeles and San Francisco.

While inland experience indicates that only one in 100 exploration targets will lead to production, those that do will take between seven to ten years from the initial exploration to full production. This is exclusive of delays caused by leasing procedures and environmental considerations. It consists of initial market studies and prospecting activities, detailed follow-up exploration and marketing arrangements, and construction of the mining system, shoreside processing facility, and market distribution system.

In this case, firm X studies the metropolitan markets appearing to have offshore resource potential. It also holds preliminary discussions with the major firms that market construction aggregate in each area to assess their receptiveness to the notion of purchasing a supply of offshore material. A decision is reached as to the prime target area.
(Time = 1/3 yr)

After leasing arrangements are made, prospecting is begun. Offshore deposits of sand and gravel are of two principal types: glacial deposits of the Pleistocene age, formed when the sea level was lower, and deposits derived from rivers that drain the adjoining land masses. In addition, submerged ancient beach deposits are known to exist. Both bathymetry and acoustic subbottom profiling are used to discover and delineate the existence, if any, of a large, linear-tending, submerged river terrace. A few drill samples may confirm the presence of sand and gravel.

Marketing arrangements are worked out whereby firm X will furnish, subject to the favorable outcome of more detailed exploration sampling, 920 thousand metric tons (1 million tons) of an agreed-upon quality of sand and gravel each year for 10 years, commencing four years from this date.
(Time = 1/2 yr)

Delineation of the deposit is conducted next, in order to answer fully two questions: (1) does the deposit contain marketable amounts of sand and/or gravel? and (2) can the deposit be mined by state-of-art techniques? With respect to (1), at issue is a) the ratio of sand to gravel and the amount and kind of impurities; b) the relative demand for sand and gravel, which reflect the market situation in the area. In some cases the demand will be for gravel, because sand may be relatively available onshore. In other cases the demand will be for sand. In this case, the ideal dredge material is considered to be 40 percent sand and 60 percent gravel. Impurities normally consist of clay, silt, fine sand, and shells. Generally, 5 percent impurities is considered a maximum for a deposit to be mineable.

Both questions (1) and (2) above can be answered by probing the deposit with a large diameter [77 cm (30 in.)] drill sampling tool. This provides a bulk sample which, while not undisturbed, is adequate for deposit evaluation.

The size of the deposit is found to contain well in excess of the desired nine million metric tons (10 million tons), it is found to be of adequate quality, and to be amenable to state-of-art mining techniques. (Time = 1 yr)

Based on the characteristics of the deposit, capital is committed for the purpose of constructing a mining system. At the same time, the marketing partner commits capital for the construction of a shoreside processing facility and purchase of trucks for the short haul to urban construction sites. (Time = 3 yrs)
[Total time to production = 5 to 5-1/2 yrs]

Sand and gravel are economically mined from the seafloor in two ways: suction dredging and clam-shell dredging. European operations favor the former; Japanese, the latter. The clam-shell is used by the Japanese because it is economic for their numerous, low-volume, close-to-shore operations. Coastal erosion problems, caused by mining too close to shore, have focused firm X's attention on a target far from shore, amenable to suction dredging.

There are two main systems of suction dredging applicable to the United States outer continental shelf: trailing suction hopper dredge and cutterhead pipeline dredge.

The cutterhead pipeline dredge has the potential to operate up to a few kilometers from shore and is described in Case II. Firm X's deposit is 128 kilometers (80 miles) from port, so it selects the trailing suction hopper dredge.

The trailing suction hopper dredge ranges in size from about 920 to about 9,000 metric tons (1,000 to about 10,000 tons). One or more high-head centrifugal pumps are used to dredge a slurry of solids from the seafloor through suction pipes. Dredging to about 37 meters (120 ft) below the water surface is commonplace; below that, jet assistance is utilized.

If the dredge is equipped with a swell compensator and articulated dredge pipe(s), most of the relative motion between the ship and the dredgehead can be accommodated. This prevents the transfer of the weight of the ship to the dredgehead and enables the dredge to work routinely in seas up to 2 meters (6 ft). An operator can assist the swell compensator in heavy seas by manually working the main winch controlling the dredge pipe, enabling the dredge to work in seas up to 6 meters (20 ft), although this is not commonplace.

The slurry, containing about 10 percent solids, is fed to the hopper(s) where most of the solids remain. The excess water flows overboard, along with the suspended fine particles. The dredge mines while in motion, creating, as shown in Figure 4, numerous shallow trenches in the seafloor, each about 1 meter (3 ft) wide and 30 cm (1 ft) deep.

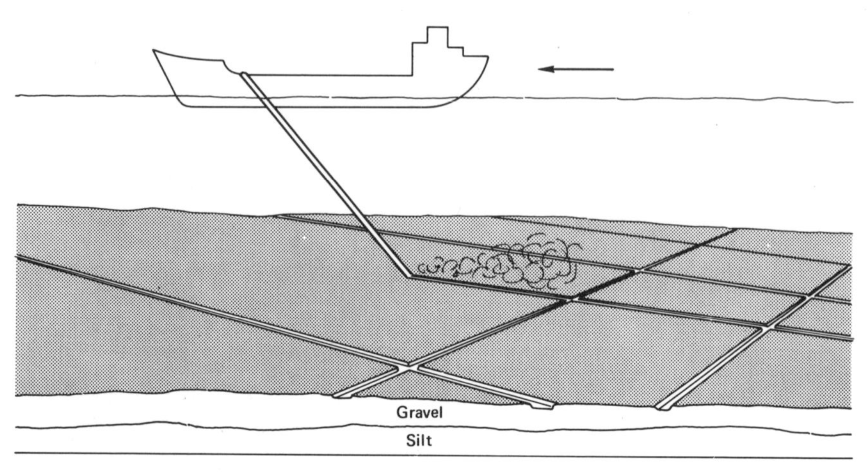

FIGURE 4. Trailer dredging.

The size of the dredge is optimized on the basis of the market arrangement, whereby 920,000 metric tons (1 million tons) per year can be utilized over ten years. Firm X decides to construct a 3,000 ton dredge and supply the one million tons per year commitment by using it virtually full-time.

Firm X estimates that it can deliver the aggregate for 50 cents per ton. Its partner will pay $1.00 per ton, process the sand and gravel at a cost of $.50 per ton, and deliver the materials to the market place for $2.50 per ton. While the United States average cost of inland-processed sand and gravel is about one-half this, the high cost of truck delivery makes this figure highly competitive in certain urban markets.

The size of the deposit may be a function of market considerations, where the market is restricted. After the economics of different systems are costed and compared, the resultant amortization period is translated into tonnage requirements. This requirement may be met by one or more deposits. The minimum required mineable tonnage is related to the marketplace and the mining systems. The deposit discovered by firm X is under 28 meters (90 ft) of water and contains zones of low-quality material. Nonetheless, the volume of high-quality material within reach of the 36 meter (120 ft) dredge contains well in excess of the required nine million metric tons (10 million tons). (Figure 5)

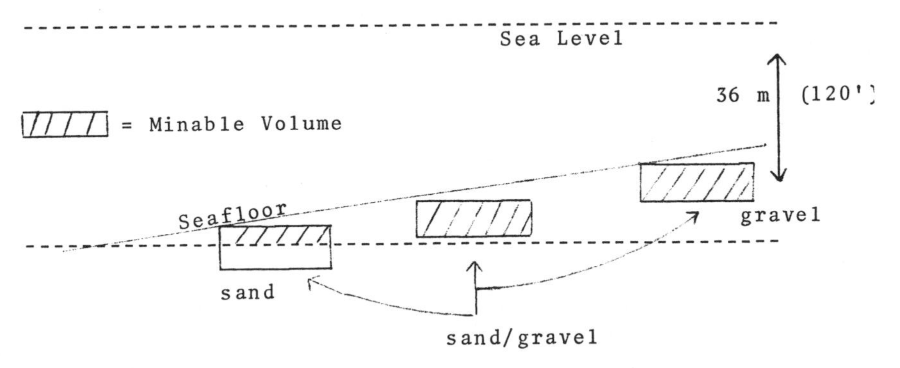

FIGURE 5

Firm X's deposit lies 96 kilometers (60 miles) offshore, and 128 kilometers (80 miles) from the processing site. (Figure 6)

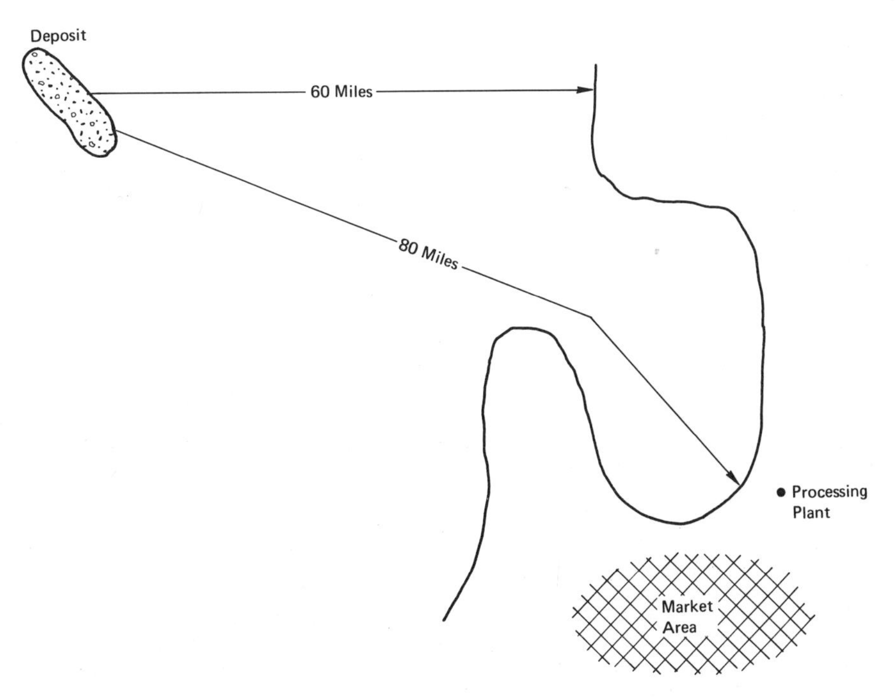

FIGURE 6. Deposit site to shore (60 miles); deposit site to processing plant (80 miles).

The mining cycle takes about 24 hours to complete, including five hours for loading at the mine site and transit to the processing plant, and three hours for unloading and return to the mine site. Operation is continuous except for about 10-15 percent down-time for maintenance and adverse weather.

The dredge utilizes a coarse-grid steel framework across the opening of the suction head to prevent large rocks from passing up the suction pipe. Coarser sizes are screened off and rejected after passing through the pump. At the other end of the particle size spectrum, fine material is washed overboard. Vibrating screens allow part of the sand fraction to be dumped back into the ocean because the ratio of sand to gravel mined is about 70:30, while the market mix requires about 40:60.

The shore-based support facility for the dredge includes wharves, stockpiling, and processing facilities, as well as a treatment plant. Dry discharging of the dredge is accomplished by scraper-buckets coupled with over-the-side conveyor belts. With this system, scraper-buckets are rapidly hauled up ramps at the forward part of the hopper and then emptied into an elevated hopper, which feeds an over-the-side conveyor belt carrying the material ashore.

Clay, salt, shells and other impurities are washed out by shoreside processing techniques, which then separate the material into a variety of sizes for blending as required by the market. A simplified flow sheet follows:

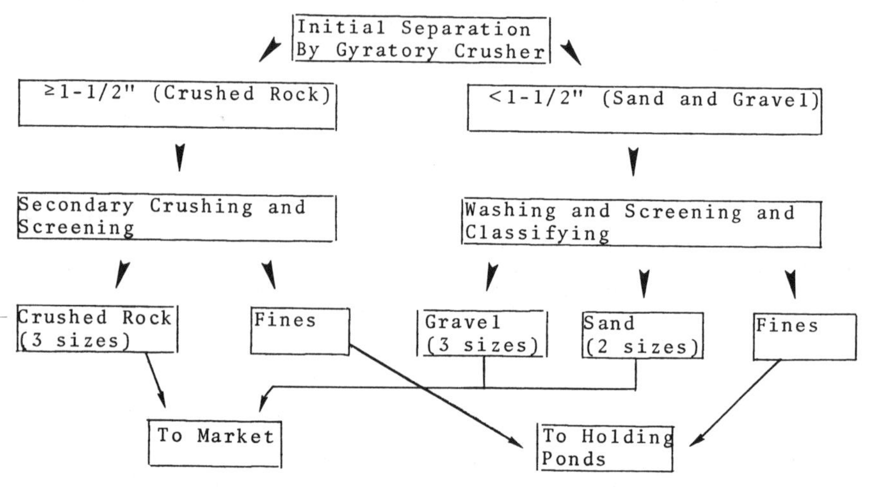

FIGURE 7. Shoreside processing.

The environmental impact potential of first order effects (e.g., excavation, turbidity plume) is obvious. In many cases the second and third order effects are easily deduced. In others, research is required to assess the ultimate consequences.

Major environmental effects include the excavation of trenches in the seafloor (complicated by rejection of non-marketable sand fraction), creation of a turbidity plume by washing overboard clay particles taken up in the dredge pipe, and production of a blanket of fines covering the seafloor downcurrent from the dredge.

Excavation volume includes the sum of the marketable material, the fraction to be rejected, and the overburden initially stripped away. In moving 920,000 metric tons (1 million tons) of product to shore annually, an equal amount of sand is rejected in order to reduce the sand: gravel ratio from 70:30 to 40:60. The 1,800,000 metric tons (2 million tons) of annual excavation results in one or more depressions in the ocean floor equal to the thickness of the deposit [3 meters (10 ft)] and equal in area to about 405 hectares (100 acres). As mining proceeds, the rejected sand released above the mined-out area partly smooths the bottom contours reducing the excavation depressions.

The turbidity plume is caused by the daily five hour excavation period that recovers 2,700 metric tons (3,000 tons) of marketable product while an equal volume of sand is obtained, screened, and rejected. The plume results from the washing overboard of the fine material suspended in the discharge of seawater from the hopper. The daily new increment of plume consists of 55,000 metric tons (60,000 tons) of water and about 180 metric tons (200 tons) of material finer than 200 mesh. Because the fines stay in suspension for days, the impact of daily activity is cumulative.

The blanket of fines settles gradually to the seafloor as it travels with the currents, building up a very thin veneer over a large area.

Firm X monitors the current pattern in the area and is quite certain where the fines are traveling. When the direction of travel at a certain time of year poses a problem to spawning grounds, for instance, firm X delays mining until the situation changes.

Shell (Case II)

Ancient reefs of oyster skeletons frequently covered by several feet of sediment, are being mined in several locations off the United States coast--all in state waters, and all, at present, in estuaries leading to the Gulf of Mexico. The dredged material is used for road fill and as a source of calcium carbonate in the manufacture of cement.

While the potential for shell mining on the outer continental shelf appears modest, deposits have been reported that could serve as important additions to the resource base in a few areas of the southeastern United States.

Generally, either a suction hopper dredge is used (as described in Case I) or a cutterhead pipeline dredge. In addition, small deposits often are mined by barge-mounted clamshells or drag-lines. The following pages describe a hypothetical but possibly typical operation on the outer continental shelf.

Inasmuch as shells are a low-cost bulk commodity of value to nearby coastal metropolitan areas, the sequence of events, as well as the time frame from market study to production, is similar to that for sand and gravel, as described in Case I.

In Case II, living reefs are found close enough to Firm X's deposit to cause concern that a turbidity plume could affect them. Therefore, a cutterhead pipeline dredge is selected so that the disposal of fines can be accomplished onshore. The cutterhead pipeline dredge is similar to the suction dredge but is equipped with a rotating cutter surrounding the intake end of the suction pipe. This cutter loosens the material which is then sucked through the dredging pumps to screens where the shells are separated from the impurities. The shells are loaded by conveyor belts into boxes while the waste impurities are pumped ashore through a floating pipeline leading to a disposal area.

The rationale presented in Case I for deposit size is generally applicable to shell deposits also. The deposit is 19 kilometers (11 miles) offshore. A fleet of barges transports the shells 22 kilometers (14 miles) from the dredge to a cement plant, while a floating pipeline transports the waste fines to a diked disposal area 26 kilometers (16 miles) from the dredge (see Figure 8). Booster pumps are stationed every 900 meters (3,000 ft) along the pipeline to keep the sediments moving in suspension.

The production cycle differs from Case I in two major ways: dredging is continuous, except for down-time due to bad weather and maintenance work, and the waste fines are handled onshore. Each barge is loaded in two hours off-loaded in two hours, and requires two hours for transit each way, for a cycle time of eight hours. Five barges, including one standby unit, are utilized to maintain the flow of shells to the cement plant. The bottom areas and excavation volumes involved are similar to those of Case I except that all excavated material travels to shore.

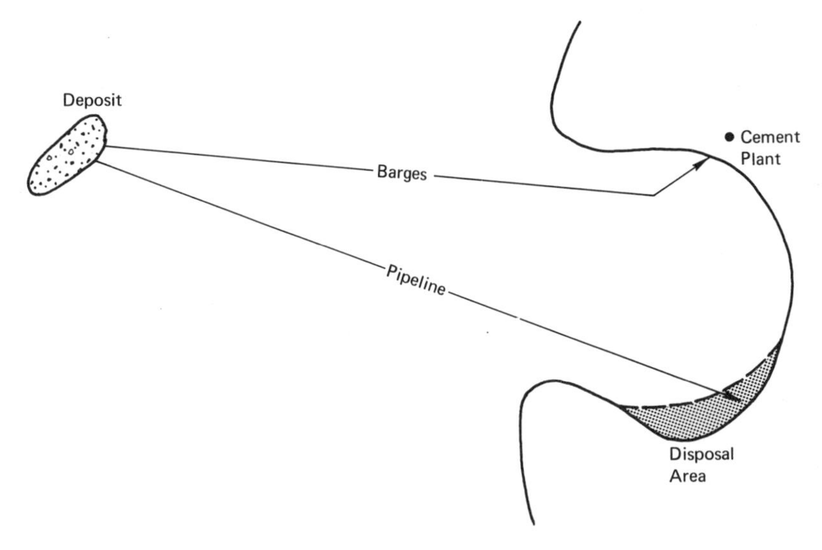

FIGURE 8. In a shell mining operation barges transport shells from dredge to cement plant while a floating pipeline transports the waste fines from dredge to disposal area.

A fill site encompassing an area of 81 hectares (200 acres) is enclosed by a riprap dike extending three meters (10 ft) above high water. The floating pipeline discharges water and suspended fine material into the impounded area. Once inside the impoundment, the solid material settles and the displaced water floats to the sea through spillways in the dike. The disposal site is designed in such a manner that nearly all suspended fines are trapped and the discharge water contains little or no more solids per unit volume than the receiving waters.

There are two main first-order environmental considerations: excavation of the seafloor and the discharge of an enormous volume of saline water into a nearshore, brackish area via an impoundment basin. The 2700 metric ton (3,000 ton) per day operation yields about ten times that amount of water. In addition, the shell delivered to the cement plant is washed and the resultant high-salinity water discharged into a nearshore area.

Phosphorite (Case III)

Submarine phosphorite occurs in several forms--as nodules, as phosphatic sands and muds, and as beds of consolidated sediments in consolidated rock. It has been found on the continental shelves and slopes of several areas of the world including Southern California (Figure 9), the south-

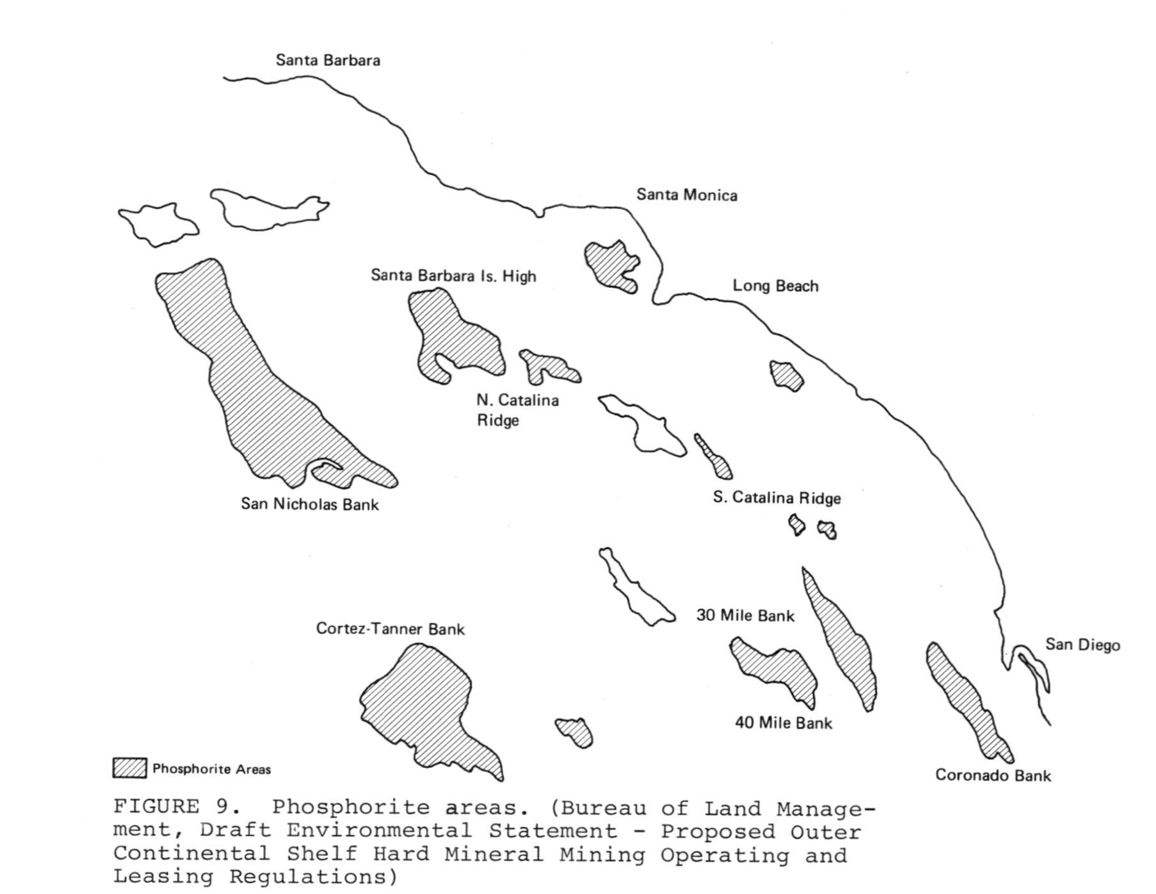

FIGURE 9. Phosphorite areas. (Bureau of Land Management, Draft Environmental Statement - Proposed Outer Continental Shelf Hard Mineral Mining Operating and Leasing Regulations)

eastern United States from North Carolina to southern Florida, Mexico, South Africa and Australia.

While a worldwide shortage of phosphorite does not exist, environmental concerns regarding strip mining on land in the United States and rapidly evolving needs , particularly for the production of fertilizer, have resulted in a strong commercial interest.

Because the economical mining of submarine phosphorite has not yet been accomplished, the hypothetical but typical mining situation described on the following pages is based solely on concepts.

The market study covers the western United States and Canada. Demand projections are good in one growth area, where fertilizer producers are interested in a new source of phosphorite rock to satisfy a part of their future needs. (Time = 1 yr).

The offshore prospecting area is explored by means of closed circuit television and bottom photography. Areas of dense nodule coverage, potentially amenable to mining, are sampled by means of drag dredge and box corer. A target area is located and tentative market arrangements are worked out. (Time = 1 yr).

A lease is secured and detailed delineation of the mineral deposit is conducted in order to assess the quantity, distribution, and quality of the ore body. (Time = 2 yrs).

The ore body appears to be promising economically, so market arrangements are completed and capital committed for the construction of a mining system, and transportation from shore to fertilizer plant is arranged. (Time = 3 yrs).
[Total time to production = 7 yrs].

Mining systems with possible application to nodular phosphorite recovery include the suction dredge, either trailing or anchored, the continuous line bucket (see Chapter 4) and the drag-dredge. Firm X selects a large drag-dredge, its size tailored to the characteristics of the deposit and the market. The phosphorite is recovered by a large dredge bucket that scrapes nodules from the surface of the deposit and feeds them into a barge for transportation to shore (Figure 10).

The deposit is found in 180 meters (600 ft) of water, 48 kilometers (30 miles) from shore, and contains 2.7 million metric tons (3 million tons) of mineable nodules of phosphorite at an average concentration of 96 kg/m^2 (20 lbs/ft^2). The dredged material is barged 64 kilometers (40 miles) to a shoreside processing plant (Figure 11) where it is washed free of impurities and loaded in railroad cars for shipment 640 kilometers (400 miles) to the inland chemical fertilizer plant.

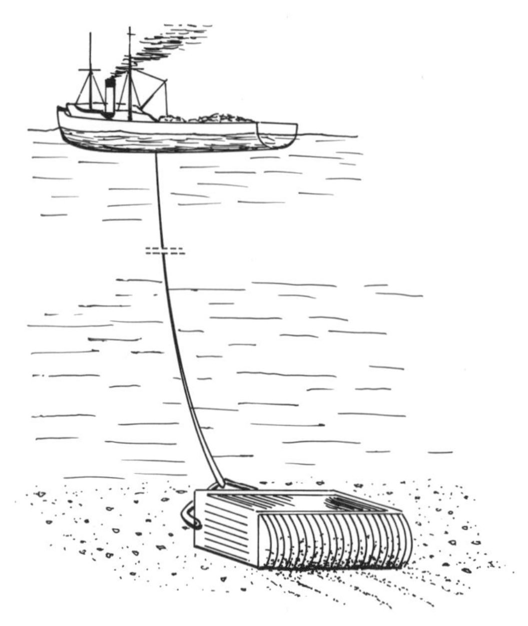

FIGURE 10. The deep-sea drag dredge. In shallow water an effective method of mining sea-floor surficial sediments. (John Mero, Mineral Resources of the Sea)

For a capital cost of $3.5 million and operating costs of $1.8 million per year, the mineral product is delivered to the railroad at a cost of $8 per ton and sold to the fertilizer plant for $12 per ton.

The annual production of 360,000 metric tons (400,000 tons) of nodular phosphorite are recovered by means of 38-cubic meter (50 cubic yard) dredge buckets, scraping an average of 18 metric tons (20 tons) of nodules from the surface of the deposit every 20 minutes. The operation is continuous except for 20 percent down-time for maintenance and weather.

The nodules are loaded into a 650 metric ton (700 ton) barge, which takes about a half-day to fill. When full, the barge steams to the processing plant (4 hrs) for unloading (4 hrs). Three barges, including one standby barge, service the production unit.

As the nodules are loaded into the barge, they are washed to remove clay and other impurities. After dry unloading from the barge, the nodules are washed again and then dried for rail shipment to the fertilizer plant.

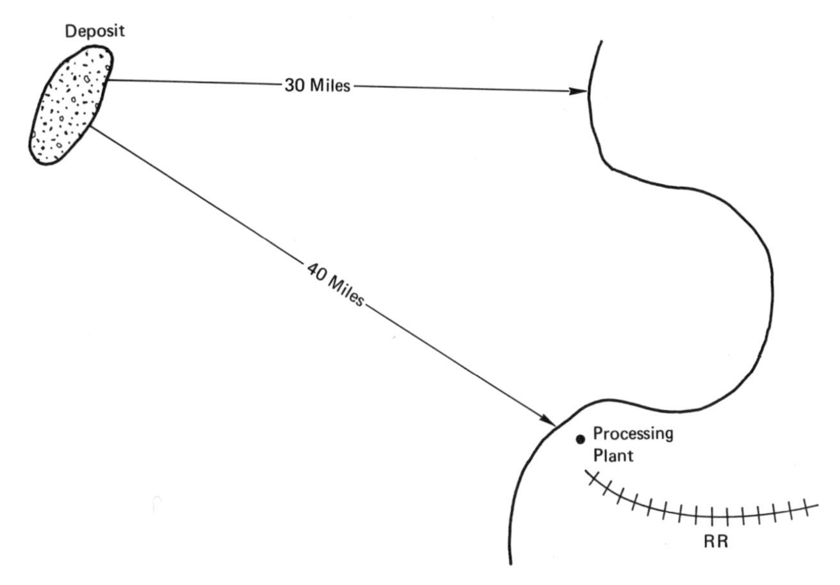

FIGURE 11. In a phosphorite mining operation the dredged material is barged to a shoreside processing plant, then loaded in railroad cars for shipment to inland chemical fertilizer plant.

The first order environmental effects are three: stripping of the nodules from the seafloor; the creation of a turbidity plume at the barge by the washing of the mined material; and the shoreside washing of the material for final removal of impurities.

At an average of 96 kg/m^2 (20 lbs/sq ft), nearly 4 square kilometers (2 sq miles) of seafloor is stripped each year to produce the 360,000 metric tons (400,000 tons). The turbidity plume at the dredge is minimal, as most of the fine material is washed out during the act of drag dredging. The discharge waste water at the shoreside processing plant will contain few fines but it will contain salt.

Heavy Minerals (Case IV)

This hypothetical operation is the mining of placer gold in the Bering Sea 10 kilometers (6 miles) offshore of Nome, Alaska. The deposit covers approximately 8 square kilometers (3 sq miles), occurs at two parallel submerged beach placers about 92 meters (100 yds) wide, 1 kilometer (.5 mile) apart, and extends parallel to the existing shoreline for almost 10 kilometers (6 miles). The water depth varies between 18 to 23 meters (59 to 75 ft) over the deposit, and the depth from seabed to bedrock varies from 9 to 15 meters (29 to 50 ft).

The gold occurs near the bedrock in a lenticular section varying between 1 and 2 meters (2 and 6 ft) in thickness. Overburden is partially cemented gravelly material with a top layer of mobile muddy sand. The ore averages 3 grams (1 oz) per cubic meter (1 cubic yd), or about 3 ppm. About 5 cubic meters (7 cubic yds) of overburden must be removed for every 1 cubic meter (1 cubic yd) of ore.

Working time is limited to approximately 185 days because of winter ice and occasional summer storms. The equipment chosen to work this comparatively low grade deposit ($2 per cubic yd at $140 per oz gold) is a 2 cubic meter (2 cubic yd) ladder dredge, capable of digging to 49 meters (160 ft) below water surface. The throughput of this dredge is approximately 31,000 cubic meters (40,000 cubic yds) per day, or 6 million cubic meters (8 million cubic yds) per work year. This capacity is necessary to reduce the unit cost of mined ore, taking into account the restraints of working time, difficult digging, and rough seas, all of which require special design features on the dredge. The ore is processed on the dredge by standard gravity methods. Concentrates, consisting mainly of black sand, amount to approximately 100 parts per million of the throughput or about 3 cubic meters (4 cubic yds) per day. The remaining 30,995 cubic meters (39,996 cubic yds) is returned to the seafloor, far enough from the working face to avoid being dredged a second time. The standard method of separation of fine gold from concentrates is by amalgamation with mercury, followed by distillation, or by differential solution of the gold in potassium cyanide, followed by re-precipitation using zinc metal as the agent. The small volume involved, when dealing with concentrates only, permits close control of these operations within a closed cycle system.

Environmental impacts result from the disturbance of the ground by the excavation process and from disposal of most of the excavated material. Fine sediments are dispersed in a turbidity plume during this period, according to the local dynamic patterns. Although a large excavation is required in the working area of the dredge, the net result of the operation consists of an increase in elevation of the seafloor by as much as 8 meters (25 ft). This could have a measurable effect on wave diffraction and sediment transport patterns in the local area.

Effects on indigenous biotic species probably result from the disruption of the benthic habitat and the dispersal of sediments in the water and on the seafloor.

Underground Mining (Case V)

This hypothetical operation is concerned with the mining of an underground coal deposit off the New England coast.

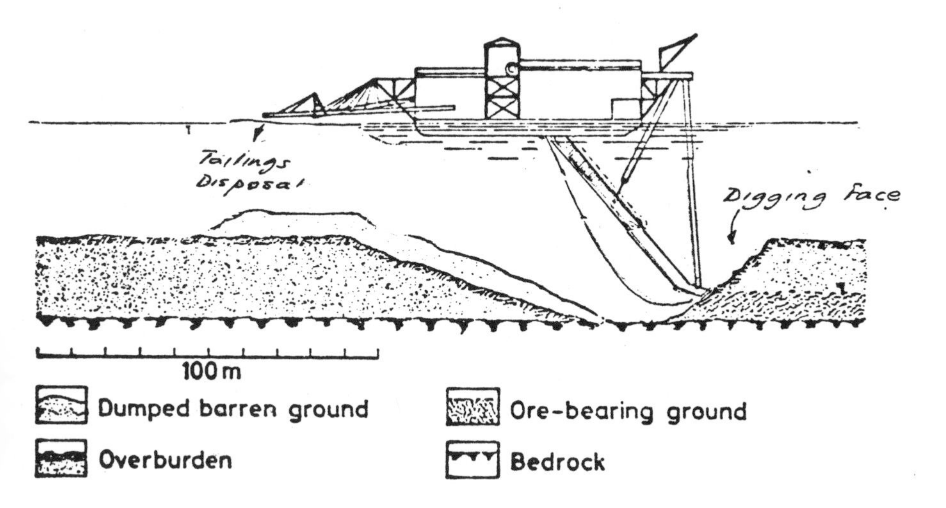

FIGURE 12. Sketch of start-up in new area.
(P.H.A. Zaalberg, Offshore In-Dredging in Indonesia)

The deposit covers approximately 307 square kilometers (120 sq miles) and underlies the seafloor by some 610 meters (2000 ft) in water depths of from 31-92 meters (100-300 ft). The average distance from shore is 11 kilometers (7 miles).

Delineation of the deposit, which was discovered by drilling into a favorable geologic area, will take several years. The number of holes required may vary according to the information accumulated, but in this case it is assumed to be 120. These will be drilled from platforms or drilling vessels similar to those used in the oil industry. The mining of this deposit, which contains a cumulative 9 meters (30 ft) of high quality coal in three generally horizontal seams, will be carried out by sinking twin shoreside shafts and driving twin tunnels to the main coal bed. (Figure 13). An artificial island will be constructed in a median area of the leases for a ventilation and access shaft. The water depth in the selected site is 37 meters (120 ft). The mining system will incorporate a standard room and pillar extraction method.

Investment costs for such an operation come to several hundred million dollars. Potential environmental impacts in the marine area will be confined to the exploratory drilling operations and to the construction of the artificial island. The impacts of the drilling will be largely confined to the drilling period during the presence of the platform. If the platform sits on the bottom there will be a limited disturbance of the seafloor. If it is anchored, minor disturbances due to the anchoring system will occur. The displacement of drill cuttings and

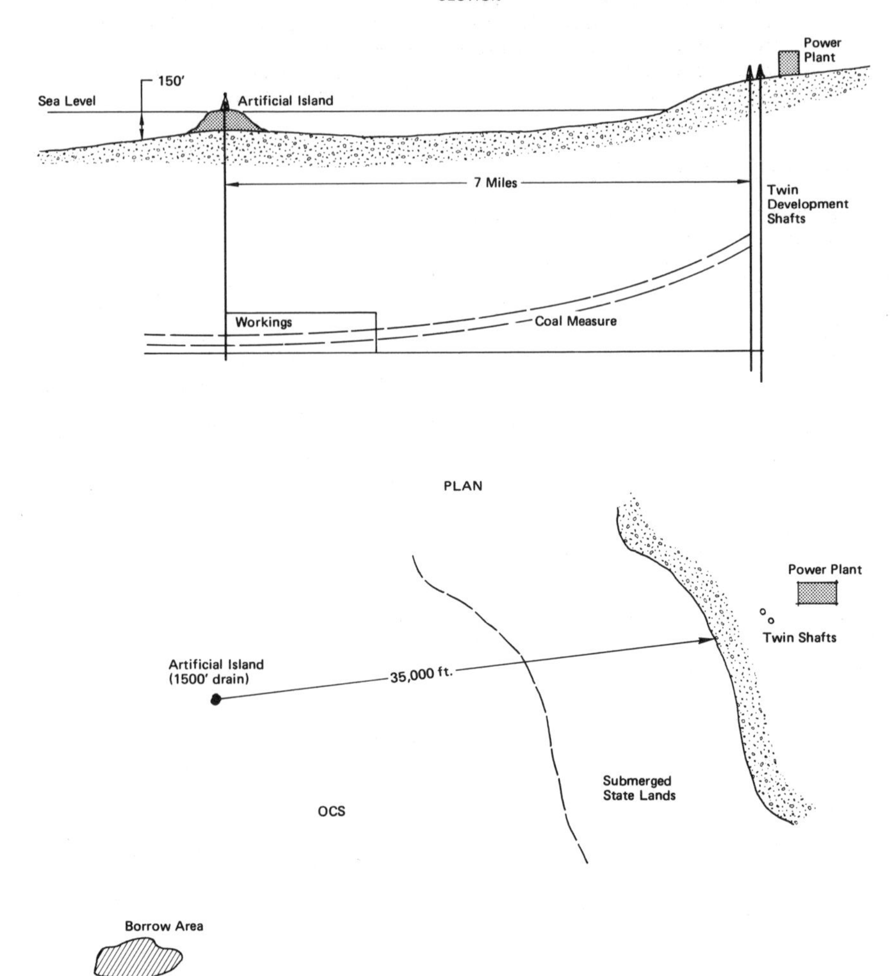

FIGURE 13. Section and plan sketches of coal mining operation.

addition of drilling mud may leave a residual of up to 91 metric tons (100 tons) of inert granular material on the seafloor around each hole. The island might require as much as 12 million cubic meters (15 million cubic yds) of fill, which would be sought from a nearby suitable marine source. The impact of the island construction might be greatest in the fill source area and would be similar to a large scale sand and gravel operation as previously described. Impacts due to the presence of the island might include diverting the current flow, altering local erosional/depositional patterns and influencing the available bio-habitats in the immediate area. Shoreside impacts are not developed here, but would be similar to a large underground coal mining operation. If the coal were to be utilized for power plant feed, it could be crushed and pipelined to any reasonable location inland.

Solution Mining (Case VI)

The hypothetical solution mining operation centers around a massive high grade copper/nickel sulphide ore deposit located at a depth of 920 meters (3000 ft) in intrusive rocks, 48 kilometers (30 miles) offshore of the southeast coast of Alaska. Water depths vary from 55 to 182 meters (180 to 600 ft). Exploration of the deposit over the past three years has been sufficient to delineate a 30 year supply of ore [at a rate equivalent to the mining of five million metric tons (5 million tons) of 1 percent sulphides per year by normal underground methods.] Controlled fracturing of the ore body in place is accomplished by drilling a pattern of 180 holes over an 8 square kilometer (3 sq mile) area and initiating a series of controlled blasts throughout the pattern, followed by hydraulic fracturing. The holes are drilled in similar fashion to those required for oil and gas production. They are cased and fitted with seafloor completion control systems. The depth of the holes drilled is between 1820 and 2450 meters (6000 and 8000 ft). Fracturing is confined to the location of the sulphide ore body, contained between the 1520 and 2450 meter (5000 and 8000 ft) horizons. No measurable ground movement is transmitted to the seafloor by this operation. Extraction of the sulphides is accomplished by pumping an acid solvent through the fractured ore body and back to the plant on land, where the metals are recovered from the solution. The major capital works involved are the pumping station, the 48 kilometers (30 mile) pipeline complex to and from the deposit, offshore facilities, the orebody injection control system, and the hydrometallurgical plant. Both the pumping station and the plant are located on shore and are served by a limited capacity marine facility owned and operated by the company. A small community of a few hundred people, mostly employees, is nearby.

Throughput of the system is around 13 million liters (3 million gals) per day. The materials cycled in the system are water, acid, iron, and precipitate copper. The active ingredients are recycled in the plant and the only toxic wastes produced are in solid form. They are shipped out for safe disposal at an approved site (probably on land).

The cost of development for this operation, including some $80 million for the plant, is $200 million to $400 million. Development drilling, pipeline installation, and plant construction from concept to completion covers a four to five year period. The *in situ* extractive system

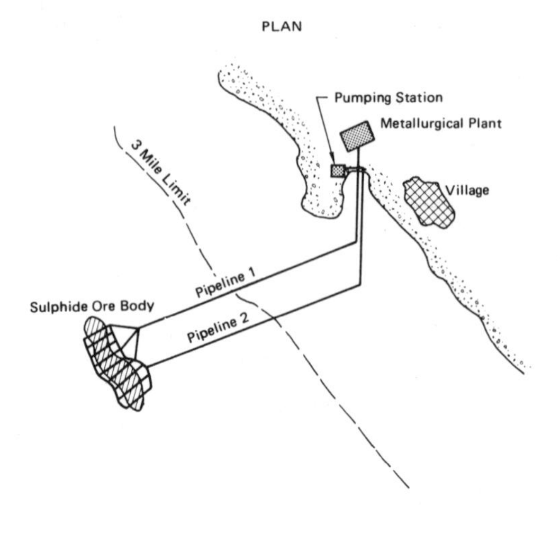

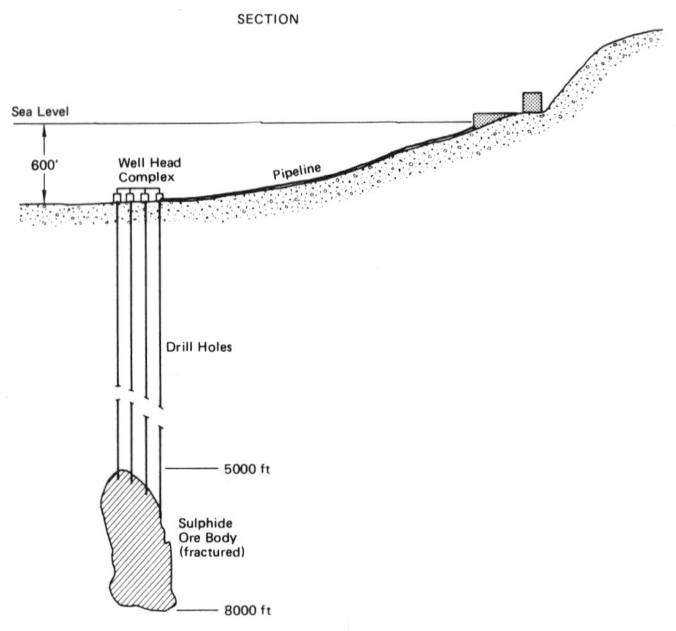

FIGURE 14. Plan and section sketches of <u>in-situ</u> methods of mining.

established here has been well developed in similar operations on land.

Environmental impacts that might occur from this operation may be surmised. During the development period, the impact of drilling will be similar to that for any production drilling program of this nature and will include the possibilities of mud or fuel spillage, some seafloor disturbance and deposition of materials from the hole. Similarly, disturbances of the seafloor may be occasioned during the installation of the pipelines and the injection complex. These structures will project in most cases from the seafloor but they can be buried or protected by smooth surfaced housings. Possible impacts of the operations are due mostly to the chance of accidental spillage, resulting from the fracture of the pipeline or injection system by an earthquake or collision. The seafloor system, which could drain in the event of such an accident at the deepest well head, assuming 15-inch lines, will contain around 17 million liters (4 million gals) of solution. All toxic solutions within the plant are recycled, but the disposal of toxic solid wastes from the shore plant was not considered in this study.

ENVIRONMENTAL PROTECTION AND SAFETY

The development of the outer continental shelf industry comes at a time of increasing concern for environmental quality. In response to this concern, United States legislative mandates require that the environmental impacts of major new public and private projects should be assessed.

The National Environmental Policy Act of 1969 (NEPA)[17] has been the single most important instrument leading to, or forcing, the acknowledgment for equal considerations of environmental quality along with socio-economic need.

The principal goals of the process of environmental impact assessments include:

1. Analysis of the adverse and beneficial, direct and indirect, environmental impacts of a project;

2. Giving the public a voice in the decision-making process;

3. Allowing environmental quality to be considered along with economic and technological feasibility; and

4. Providing a stimulus to industry to conduct its activities in an environmentally safe manner.

There is a need for a careful environmental assessment of the new mining industry. Such an assessment is required by NEPA and subsequent implementation guidelines.

More importantly, a new industry is being developed in an environment that is still poorly understood.

Besides the requirements of NEPA, there are important reasons for considering environmental quality at the beginning of marine mining operations--namely: to allow an orderly development of the new industry while protecting or enhancing environmental quality; to facilitate continuing multiple use of the outer continental shelf; and to provide for industrial development in a stable regulatory regime.

Thus, if it is in the national interest to undertake heavy mineral mining on the outer continental shelf, it is also in the same interest to develop orderly procedures for the protection of the environment. These procedures must become part of the regulatory mandate so that the scope and the limitations are understood prior to the time of initial lease agreements.

The rationale for this position lies in past experience. Too often in the history of industrial society little or no notice has been taken of existing environmental conditions at the outset, or of subsequent alterations in the environment until irreversible change or damage has occurred.

In general, private enterprise potentially involved in outer continental shelf mining is aware of the mandate for protecting environmental quality and is ready to comply with reasonable constraints that provide for protection of proprietary technology and do not require monetary outlays so great that inadequate compensation for capital and operation investments results.

State of Information

In order for environmental assessment to be completed for mining the outer continental shelf, the first step must be to characterize ambient conditions at the mining site. Following this characterization, the specific activities, including any discharges that may affect the environment, must be considered. Using this information, the environmental impact of an outer continental shelf mining operation can be assessed.

The diversity of nearshore physical environments and the nutrient input from land sources combine to produce, on the shelf, the greatest biological diversity of any area in the ocean. Knowledge of the existing physical, chemical, and biological environmental conditions is most extensive for the nearshore continental shelf margins. But even that knowledge can be grossly inadequate, as has been shown in a number of instances. For example, prior to the Santa Barbara oil blowout in 1969, no published biological survey had been done in that area since the

1956-1959 sampling by the Allan Hancock Foundation (of the University of Southern California) for the California Water Quality Control Board.[18,19] Hence, it was impossible to document the impact of the spill on prior existing conditions and compare the recovery with those conditions. All that could be documented was the improvement (or recovery) that followed the worst period of the blowout. This is not a situation unique to Southern California, by any means.

As mentioned earlier, the first step in gathering information is to turn to the available literature, seeking available data and other information about the specific area to be developed and its adjacent areas of influence. The immensity of ocean areas, compared with the relatively few environmental studies already existing, make published sources a poor supplier of information in most instances. Further, the limitations of data collected on worldwide cruises are severe in terms of area covered, time spent at individual stations, and collection methods.

In many instances, specialists were not available to sort and study the vast amounts of data gathered, and years passed before the results were published. Present-day cruise collections are stockpiled in quantity at the Smithsonian Sorting Center, awaiting funds and specialists to be able to identify specimens and interpret data. No doubt considerable environmental data exists as the unpublished property of industrial concerns, but these are not available as yet to the scientific community as a whole.

The output of recent worldwide studies pertaining to the continental shelf is relevant. During the course of this study a search of Biological Abstracts and Ocean Abstracts for the past few years located approximately 200 titles. Of these, about a third were biological, each concerning mainly a single group of animals. The Foraminifera were most prominent because of their application in locating oil. Sedimentology and mineral resources dominated the other citations; few classical oceanographic papers were included.

A computer search keyed to outer continental shelf resources was initiated through the National Technical Information Service (NTIS). Again about 200 references were abstracted. Many of these related to economic and legal questions. Very few could qualify as assisting in a baseline study of a proposed lease site or of a larger area. This was by no means an exhaustive search of the literature, but it seemed to give an overview.

The need for baseline oceanographic information follows from the concept that the natural environment functions as an ecosystem. Biological equilibrium does not truly exist in the environment if very lengthy periods are involved, but the natural system, in a shorter time span, will provide for balanced feedback between inorganic and organic living cycles. Although natural cycles may shift, as they do in long term shifts in thermal regimes or in evolutionary genetic drift, they remain fairly well stabilized overall.

Existing environmental knowledge of the outer continental shelf is much more limited in scope than is that of the inner shelf because of greater accessibility to the inshore region by smaller boats and the less rigorous demands on sampling gear in shallower waters. In this context, it is noted that the endurance and reliability of oceanographic equipment is generally poor, and uniform calibration standards are lacking for the majority of oceanographic instruments.[20]

Geophysical and oceanographic data have been collected in seas adjacent to the continental United States by a relatively few institutions and expeditions. In 1961 and 1963 the National Oceanographic Data Center published listings of 342 oceanographic vessels in operation throughout the world.[21] Of 160 U.S. vessels listed, 116 were large enough to be equipped with oceanographic instruments. Educational institutions operated 33 of these. In 1974, the University National Oceanographic Laboratory System (UNOLS) listed only 24 oceanographic vessels operated by universities.

There is a need to increase the support for oceanographic research in the United States if outer continental shelf resources are to be safely developed and managed. At present, there is little federally sponsored research on the outer continental shelf relevant to mining, though the Army Corps of Engineers and the Office of Sea Grant support modest programs.

The teaching of organismic biology has declined to the point were many biology students never see whole animals and cannot identify even common local organisms. Now, molecular biology has captivated so many academic institutions that there are relatively few experts who are competent to carry out baseline studies and to evaluate the ecological system either locally or worldwide. New approaches to teaching such work are needed if baseline studies are to have any real value. The fact that jobs are now available for competent organic biologists has already given impetus to a return to teaching about whole plants and animals.

Because published information regarding the outer continental shelf environment is scarce, the existing environmental conditions at a potential mining site must be nearly always characterized through original baseline surveys. However, the scope and extent of an adequate baseline study has not been defined and accepted as valid by the scientific community. Failure to compile adequate baseline data prior to the installation of industrial facilities occurred most often in the years prior to NEPA. It continues to occur in numerous instances where lack of scientific judgment combines with poor funding and lack of time to produce superficial results. Furthermore, regulatory agencies have been poorly staffed, especially in view of the large number of impact reports and statements now required, to give more than cursory attention to the preparation and review of environmental impact statements.

Thus, adequate baseline surveys must be defined. A baseline should have at least quarterly samplings from a specified pattern of locations over at least one annual cycle. In other words, baselines are both time-and location-dependent.

Industrial organizations and public agencies must be willing to provide adequate time lines and funds for a reasonable survey and analysis. It is not enough to survey most Pacific and Atlantic continental shelf areas a single time, especially if the only convenient time for the agency is between October and April. Finally, resource commitments, in terms of manpower, shiptime, and funding, must be expanded to accomplish this large task.

While knowledge of the ambient environment of the outer continental shelf is severely limited, even less is known of the mining activities that may produce changes in the environment and the extent of those effects. While various types of offshore mining exist in foreign countries, no mining of the outer continental shelf has occurred to date in the United States. Although certain parallels exist between navigation channel dredging and outer continental shelf mining of sand, gravel or phosphorite, there are many differences. The physical and chemical properties and biological communities in nearshore waters are somewhat different than are those on the outer continental shelf. Further, sedimentary material nearshore frequently contains contaminants generated during years of industrial activity. Outer continental shelf mineral resources and their associated sediments will probably contain only natural levels of trace elements and other contaminants. Because of these limitations on knowledge, the environmental impacts of outer continental shelf mining are difficult to assess.

Planning Baseline and Monitoring Criteria

Details of parameters and techniques for measuring environmental quality in regions of the outer continental shelf are not standardized at present because past investigations have been carried out by different institutions at different times and localities. Equipment may be limited or prototype in nature rather than standardized.

New efforts have been made recently by the National Oceanic and Atmospheric Administration, Environmental Protection Agency and the Bureau of Land Management to arrive at a consensus among the scientific community as to what constitutes adequate scope and methodology for these surveys.[22,23]

Broadly categorized, these cover physical and chemical characteristics of the water and sediment and the nature of the benthic and pelagic community ecosystems. Frequency of sampling in space and in time are often more difficult problems on which to obtain a consensus. Biologists are aware of the seasonality of biotas and many consider quarterly sampling as the absolute minimum for an adequate baseline. Most are aware that longer-term shifts in water mass and thermal patterns are not delineated by a one year survey. A case in point is the northward flow of the warm Davidson undercurrent or counter-current along the southern California coast which flows in the winter months. Normally, it reaches only to about Point Conception, but in some years, warm water fauna have reached northern California and Oregon where they persisted for several seasons. Presence or absence of a given species in a limited survey area in consecutive years might have led to the conclusion that some localized impact was the cause of its disappearance if the space and time of the survey had been too limited.

Isaacs[24] has pointed out that a survey of sediment layers in the anaerobic basins off California showed that sardines historically had perhaps a 40 year local population cycle and probably had not in fact disappeared because of over-fishing or pollution. Thus, a baseline cannot provide all the answers but can certainly serve as the standard for comparison when long-term monitoring is built into the system.

Monitoring over a sufficient area during prototype leasing and during operations should provide information on both natural shifts and impact of operations.

Environmental Impact of Technology

There are two distinct categories of potential impacts from outer continental shelf mining operations. The first consists of turbidity and possible chemical shifts due to the mechanical resuspension of sediments caused by the method of gathering or draining the minerals. These mechanical processes introduce no new or unnatural molecules to the environment, but benthic organisms directly beneath the tracked vehicles or suction devices are destroyed, and other nearby organisms may be silted over and killed. Available data indicate that temporary turbidity of adjacent waters or limited sedimentation has much less impact on benthic organisms than was originally predicted.

Technology for processing minerals at sea is presently limited, but such contingencies must be considered. For example, any process that radically alters the pH or dissolved-oxygen content of receiving waters will affect biota in at least a limited area.

Natural oil can affect plant and animal life in some cases as a study of oil seeps off Santa Barbara has shown. Research on petroleum spills has demonstrated that while naturally occurring crude oil contains toxic and carcinogenic agents, it is a more dilute form than in refined oil. Marine bacteria attack these toxic chemicals.

Similarly, chemical wastes from mineral processing are expected to have toxic effects. Environmental damage may be prevented by experimental laboratory testing during the design technology phases, which might be more economical and effective than waiting for on-site monitoring to determine possible impacts.

Mining leases will need to carry stipulations regulating any on-board processing that might be developed in the future. It is recognized, what is more, that onshore processing may have great environmental impact, and eventually a choice may have to be made between the relative impacts on land or at sea. Potential onshore effects of outer continental shelf development have been considered in detail by the Council on Environmental Quality[25] and the National Ocean Policy Study.[26] These processes might be more environmentally safe at sea, even if some adverse effects occur there, than in the confines of inshore waters or on land. However, some spills of toxic chemicals may be less safe in the ocean than on land, because on land the spill could be more easily contained. Some toxic chemicals in trace quantities may be dangerous, and the ocean circulation could cause toxic agents to spread out and damage life in a large geographical area.

Bays, estuaries, and the inner continental shelf are among the most fragile ecosystems in the world. Conflicting demands for land use and the difficulty of adequately diluting wastes on land or in semi-enclosed bodies of water tend to emphasize this aspect.

Conversely, the deep ocean apparently does not involve the diversity of environments in limited areas that are characteristic of the inner continental shelf. The biota are much more limited.[27] Variations in the outer continental shelf environment are not well known but seem to be far less vulnerable to trauma than those in the inner shelf areas.

REGULATIONS AND LEASING

The shared desire of environmental interest groups and the industry is to have clearly defined environmental standards set in advance of mining. While it is difficult to predict the operating stipulations which may be written into a federal lease or license, the environmental procedures which might be followed in the orderly development of outer continental shelf mining are described below.

When certain regulations are issued by a government agency, NEPA requires that an environmental impact statement be prepared to accompany them. In February, 1974, the Department of the Interior issued Proposed Leasing Regulations for outer continental shelf hard mineral resources, in response to the increasing need for the resources, coupled with growing industry interest in mining them. As required by the National Environmental Policy Act, a draft EIS was prepared to accompany the proposed regulations.[28] This draft EIS is termed programmatic in that it considers the environmental implications of adopting the regulations rather than specific mining operations and areas. Should the recommendations of this report be followed, a different set of regulations would need to be issued. Specifically the Panel recommends that a licensing system be substituted for a leasing system. Thus, the draft EIS would either need to be rewritten or amended.

Finalization of licensing regulations and acceptance of the final EIS will be followed by industrial prospecting activities. Prospecting will identify deposits of potential economic importance.

The first procedural requirement to further the objective of environmental safety regulations would be to institute baseline studies during the preliminary prospecting phase. The scope of the baseline investigations would be extensive enough, and gather sufficient information, to permit an analysis of the area as an ecosystem.

A second environmental assessment should occur when the Department of the Interior has prepared a ten-year licensing schedule. Regional programmatic statements will be prepared, as detailed in the Regulations section of this report.

A third and final EIS would be prepared at the licensing stage for a specific area. An applicant for a production license would have to submit a work program which would include a list of the principal minerals to be found; the preliminary plans for exploration and development along with methods of mining and waste disposal, and transportation of the mineral product to shore; the possible impact of the operation on the total environment; and the measures to be used to minimize the environmental impact of the proposed operation. Using this information, together with their own independent analysis of the environmental impacts, the Department of the Interior would prepare and issue a draft EIS.

Because a nominee would be required to submit the environmental report described above, it is essential that environmental baseline studies for the proposed area be initiated during the exploration phase. While a portion of the baseline survey costs must be borne by industry, the Department of the Interior must also develop and fund an aggressive program that will characterize baseline conditions. To avoid duplication of effort, cooperative programs, guided by an evaluative organization, should be established. The character of the evaluative organization is recommended below.

The production license Environmental Impact Statement will address operational conditions described by the applicant. This EIS will therefore be specific as to site and technology to be employed and should quantitatively describe impacts. However, it appears that a detailed impact assessment cannot be accomplished with the present data base. While the existing environment will be studied prior to tract nomination, operational effects on the environment will not be known.

It is therefore prudent that licenses for prototype mining operations be authorized before full scale licensing begins. Several prototype mining operations should be authorized in sites representative of the outer continental shelf region with hard mineral potential. To gain a maximum amount and variety of knowledge from these prototype operations, it would be wise to select the sites and operational modes to represent the full spectrum of environmental impacts which will likely follow under full scale licensing.

The environmental effects of the prototype operations must be monitored thoroughly. The major environmental objectives of prototype operations are to identify the environmental effects of mining on the outer continental shelf under actual operating conditions; to allow a small number of mining operations to begin under supervised conditions rather than allowing a larger number of less well controlled operations before the environmental implications are understood; to allow mining to take place for a specified period of at least a year, so that long-term environmental impacts may be weighed against short-term changes; and to allow the mining industry to experiment with technologies in order to minimize adverse environmental impacts.

When issued a license, the licensee will be required to submit a mining plan to the Department of the Interior. This plan must be responsive to stipulations placed on the license by the federal government. Thus, the results of monitored prototype operations would find direct and immediate application. Regulatory agencies would be aware of long-term adverse effects as well as feasible methods and technologies available to mitigate them. Industry would also know what stipulations it could expect, attendant to accepting the lease.

At the early stages of full-scale industry, license stipulations would probably include a monitoring program. In areas where the substrate texture is important to the ecosystem, as in spawning areas, restoration should be required. Other stipulations may include seasonal operating restrictions, warranted by migration patterns of sensitive species, or limits on the disposal rates of particulate wastes.

It is considered essential that a scientifically knowledgeable body, independent of a licensing agency and the license holder, alike, be given the responsibility and authority to evaluate the results of the environmental baseline studies and to monitor outer continental shelf mining operations. This committee should be appointed by an independent advisory group. Their functions should include supervision of proposed scopes of work, stipulation of the level of effort required, evaluation of the level of effort required, evaluation of the results of baseline and monitoring studies, and specification of modifications, additions, deletions and restorations for mining operations. The committee should have adequate staff and funding to provide a meaningful evaluation.

Objectives and Content

In assessing the management and regulation of heavy minerals mining on the outer continental shelf, the Panel accepted the objectives laid down by the Department of the Interior. These objectives are:

1. Orderly and timely resource development, with prevention of waste in the extraction of mineral resources;

2. Protection of environmental quality and the achievement of exemplary practices in mining on the outer continental shelf;

3. Return of fair resources value to the public; and

4. Impartial application of laws, regulations, and orders to operators.

Three set of circumstances were considered by the Panel in its deliberations:

1. Hard minerals mining under the ocean is an activity for which there exists only a very limited experience base.

2. Both the resource base and the environmental data base necessary to formulate regulations are very limited.

3. Hard minerals mining under the oceans will take place in areas where interested parties reflecting quite different values will demand participation.

The achievement of the stated goals, under the circumstances identified above, requires government procedures and management arrangements that provide decisions based on uncertain data. These decisions must reflect the need for accommodation and compromise among interest groups with conflicting values and objectives. In summary, governmental management must be responsive to a changing and growing data base within a fluid political environment.

Relevance of Oil and Gas Experience

It was the Panel's initial assumption that appropriate management techniques could be obtained by extrapolating from the nation's past experience with the development of outer continental shelf oil and gas resources. Like oil and gas, the major legislative basis for governmental regulation of hard minerals mining is derived from the Outer Continental Shelf Lands Act and the NEPA Act of 1969. Several Panel members, however, raised questions concerning the similarities between offshore oil and gas development and hard minerals development. After extensive review, the Panel concluded that the differences between outer continental shelf oil and gas development and hard minerals development made it impractical to extrapolate from one regulatory system to the other. Two categories of concern led to this conclusion.

1. Hard minerals mining systems may involve collection of minerals by mining vessels by means of dredging of commodities which range from unconsolidated deposits such as surficial sand and gravel in shallow water, to manganese nodules located in great depths. Also, marine mining may eventually involve underground deposits whose development will require construction of fixed facilities and shaft sinking operations.

   The range and variety of activities was potentially so great, and the necessary technology in such an early stage of development that the oil and gas analogy was not considered useful. In this connection, it should be noted that the most advanced technology is at the present time in deep ocean mining and not shallow, near-shore mining. Unlike oil and gas, it is not at all clear at the present time that the necessary technology can evolve through a gradual movement of operations from relatively shallow water into deep parts of the ocean.

2. The social and political context in which hard minerals mining will develop is distinctly different from that in which oil and gas developed. On the one hand, there is a growing perception that the nation and the world may be facing a minerals shortage. A consequence of this may be a desire for the United States to seek maximum self sufficiency in minerals resources should foreign sources become less available. Quite the opposite situation existed when the oil and gas management system was developed. On the other hand, there is a widely manifested concern that every reasonable effort should be made to protect the environment. Responding to this concern requires the development and clear

interpretation of reliable and convincing new environmental and resource data. It also requires governmental decision-making which generates public confidence, as a result of open decision making based on publicly available data. The legislative history of NEPA suggests that open information and open decision making were primary objectives of the Act. In practice, NEPA places responsibility on government for early notification to

> interested groups of pending governmental actions and provision of the data and rationale behind decisions by government. These provisions of NEPA have added substantial responsibilities to the government and industry, in their mineral management program.

In no area are the problems posed for oil and gas more graphically illustrated than for those surrounding government access to, and handling of, resource data and interpretations developed and paid for by the industry. Critics of the government's resource management programs charge that industry has better resource information than the government manager and regulator. This has led to recent proposals by the government that industry provide all pertinent oil and gas data. Quite understandably industry resists these pressures, since these data represent potential competitive advantages in bonus or royalty bidding for leases, now required in the Outer Continental Shelf Lands Act.

Should government obtain such data under present procedures, the basic social issue is not resolved since government would apparently view it as proprietary; that is, the data could not be made public or included in environmental impact statements even if deemed appropriate. Not only would the company, at its own expense, have to provide data to the resource owner (government), but it would also have to provide data to its competitors, or other potential bidders on leases.

In summary, the Panel concluded that credible government resource management requires that government have the best available resource information, and that the public recognize that to be the case. Only public availability of resource data can achieve that objective. This objective cannot be achieved under the current system that governs oil and gas resources.

General Findings

Given the facts that both technology associated with marine mining and the social circumstances under which it will develop differ significantly from the oil and gas case, the Panel concluded that it could not design a regulatory system for mining that would achieve the nation's objectives within the constraints of the existing Outer Continental Shelf Lands Act.

The central deficiency in the present Act is its requirement that leases on hard mineral resources be allocated on the basis of competitive bids that utilize bonus payments as the single variable.

The Panel's recommendations then propose making some clear-cut and potentially controversial tradeoffs.

Specifically, the Panel recommends an approach to government regulation and management of outer continental shelf mining that exchanges early financial advantage, and administratively clean allocation of mining leases by sealed bids based on royalties, for the early and complete information advantages of a licensing system that uses work program proposals as a basis for allocation.

Where mineral development is judged to be in the national interest, the Panel believes that industry capital should be invested in exploration and development activities, not in bonus money. The other side of this is that government be put in a position to provide the best informed and most imaginative management of mineral resources. The Panel has purposefully tried to provide an arrangement under which there is no significant economic advantage to a company in retaining information as a proprietary item.

To ensure maximum social responsibility on the part of the involved government agency and the licensee, the Panel recommendations rely heavily on the requirements of NEPA.

Although the requirements under Section 102 of NEPA are still being defined in the courts, it is the intent of these interpretations, as well as the Council on Environmental Quality's guidelines, that NEPA's requirement be interpreted very broadly. Specifically, environmental impacts go considerably beyond impacts on the immediate physical and biological environment. Court opinions and CEQ guidelines have indicated that, to the extent possible, the full range of impacts on the social and economic system must be assessed. A major purpose of the environmental impact statement is to ensure provision of broad-based information to policy makers and the public. The impact statement process reflects a growing recognition of the complexity of our society and the irreversibility of many of its decisions. The impact statement should be managed as a purposeful effort to provide a process for responding to conflicting values in advance of major federal actions, rather than after the fact.

Successful political accommodation requires that the impact statement process respond to several needs. These are as follows:

1. A substantial lead time is necessary. That is, probable federal actions need to be identified far in advance of those actions being taken.

2. The government needs to insure maximum public access to the decision-making process for all interested parties. It should be recognized that consumer and environmental interest groups are frequently only alerted to actions at a relatively late stage. Thus their response is frequently to attempt to block the action, in part, because these parties are both uninformed and faced with very short time constraints, but further because their interests have not been adequately addressed in the past.

3. The government should be responsible for providing all interested parties with maximum available information. This information should be made available as early as possible, be comprehensive, and include interpretation.

   Public interest groups frequently have limited resources and inadequate interpretation capabilities. Vigorous efforts by government to provide comprehensive information will mitigate a repeated concern that selective release of information is being used to support the developer's position. The impact statement process should represent a system that provides "no surprise" for any of the parties.

4. Effective response to conflicting interests requires developing comprehensive planning along the coastline and on the outer continental shelf. A major problem with present outer continental shelf management is that each decision, each lease sale, each proposal to take action, is essentially a *de novo* proceeding. It starts from ground "zero" and requires fighting all the same issues over again. Mining may impact on the coastline in some way. Coastlines have been the predominate concern of those who can be expected to oppose ocean mining. Only with the development of coastal zone planning can the same highly charged political fights be avoided in the future.

It is the Panel's belief that only a system which responds to diverse concerns on a repeated basis can hope to meet the needs of a rapidly changing world. It has attempted to ensure a responsive technology by recommending performance standards that assume the use of the best commercially available technology. The objective of the system is to ensure that industry and the federal government make continued efforts to improve the technology, both for purposes of economic return and protection of the environment, and for other uses of the ocean.

The intent of the regulations that the Panel recommends in the following section is to create a system of procedures that will make the management of these resources responsive to a broad set of public concerns.

Regulatory Principles Covering Hard Minerals Mining on the Outer Continental Shelf

I. Prospecting (Government Permit Required)

Prospecting, using such methods as magnetic, electric, gravimetric, and acoustic surveys, as well as bottom-sampling and shallow coring, should be open to any United States citizen or company (or any citizen or company of a nation with which a reciprocity agreement has been signed) upon issuance of a permit by the Department of the Interior. (This does not apply to academic research organizations which are exempt from these provisions.) Parties wishing to carry out such prospecting should submit a proposed prospecting plan to the Department of the Interior. The plan should describe the type of prospecting anticipated, and the general area to be surveyed. It should include specific assurances that such exploration would neither damage the environment nor conflict in any significant way with other users or interests in the area. In those cases where some conflict exists, the conflict should be described for purposes of informed government decision-making. Approval of prospecting should not carry proprietary rights to any minerals discovered. The Department of the Interior should be obligated to respond to a request for a prospecting permit within a reasonable time period; a nonresponse within a given time period should be defined as a formal approval to proceed.

Rationale: The Panel's recommendation reflects two conclusions:

It is to society's advantage to encourage the most extensive and complete gathering of information on mineral resources. Although during the prospecting stage these data remain the property of the prospector, they are necessary before a company will proceed to the detailed exploration stage outlined in Principle IV.

The Panel found no reason to believe that adverse impacts will be associated with normal prospecting techniques. The assumption is that issuance of prospecting permits will generally be *pro forma*; however, a formal permit is recommended as a filter to catch those rare instances where something out of the ordinary might be associated with prospecting. The Panel believes that inaction should not be an option allowed the government resource manager as a means of blocking prospecting. Thus, the recommendation is that a long delay should be formally construed as approval.

II. Licensing Schedules

The Department of the Interior should work toward early issuance of 10-year licensing schedules.

Rationale: The Panel believes that a 10-year licensing schedule is an essential framework around which both federal and state planning must take place. This schedule should reflect areas nominated by companies based on prospecting data, and nominations by the government based on its own data. Mining activities must be integrated into coastal zone planning and such planning takes time. Further, it provides a time frame for planning by the mining industry.

Such a schedule has several other benefits. It serves as an early warning device for parties representing all vested interests. This is especially true for state agencies and local interest groups who may not normally monitor activities in Washington, D.C. Finally, inclusion of a new region on the 10-year schedule, which would be updated yearly, would act as the trigger for the preparation of a regional programmatic impact statement as covered under Principle III.

In summary, the 10-year license schedule is the first step in insuring that the long lead-times are provided. The Panel believes this to be necessary to achieve the political accommodation and data collection necessary for stable mining activity.

III. Regional Programmatic Statements

With the inclusion of any coastal region on the 10-year licensing shcedule, the Department of the Interior, in close cooperation with other concerned agencies, should prepare a programmatic impact statement. These regional programmatic statements should be general development plans for the region, including mineral, land-use and environmental concerns. The purpose should be to define the role and the relationship of hard minerals operations to the overall uses of that region. These statements should be subsidiary to the more general statements issued in advance of mining regulations and should be a vehicle for updating or modifying that general statement. Regional programmatic statements should provide early access both to information and policy-making for all interested parties, and therefore, be an early step in the process of political accommodation. All programmatic statements should have a clear-cut obligation for regional public hearings. The programmatic impact statement should specifically include an assessment of federal management capabilities with regard to these minerals.

An *ad hoc* committee should be constituted by the Council on Environmental Quality to review programmatic impact statements. The committee should represent a broad range of interests and expertise. A report of the review should be provided to the public.

Rationale: The Panel has placed great importance on the regional programmatic statement. First, it offers a way to escape reassessing the total world resource picture in each license impact statement. Second, the Panel believes that mining must be assessed in terms of its general regional impact and viewed as a regional activity. Third, these statements provide a vehicle for early involvement of all interested parties in the decision-making process. This latter includes two elements: identification of the various concerns reflected by different interests, and the collection of any appropriate information those interests may have. Regional impact statements also serve as a way to inform and provide appropriate information to interested parties. Fourth, the regional impact statements, by triggering early contacts, assure the concerned parties that they are not involved in an after-the-fact process of justifying a decision. Fifth, the statements can be used as inputs to coastal zone planning. This contributes to state planning agency needs and assures that the mining activities will have appropriate shore-line support facilities built into coastal zone plans, thereby reducing the chance of controversy over such facilities at the point they are needed. Finally, these statements provide a vehicle for early coordination among the multiple federal agencies likely to be concerned with the mining activities. A particular benefit in this connection should be to assist those agencies responsible for collecting environmental data in planning their research program.

A central problem in the past has been the lack of capability of the staff of the federal agency to manage resource activities. For this reason the Panel believes the impact statement should identify the prospective personnel needs. The best regulations and procedures have little public credibility if there is no professional staff to carry them out.

The Panel has recommended a review of the regional statements by an ad hoc committee constituted by CEQ. That recommendation reflects the Panel's conclusion that the agency preparing the statement would benefit from a review by a group of expert consultants selected by a government body reflecting environmental interests. Such a review would add to the public credibility of the decision-making process.

IV. Detailed Exploration Licensing

A. Non-Competitive Licensing

In the absence of competing requests and/or known marine mineral deposits, and upon submission of a detailed exploration program, the government should issue an exploration license. This license should allow for significant sampling, deep coring, mapping, and assessment of the tenor of the ore. The license application should require detailed descriptions of the activities to be carried out as well as the specific location as defined by coordinates. Upon completion of this detailed exploration, all data and interpretations should be made available to the Department of the Interior. Normally, these data would be made available as part of the application for a production license, both data and interpretations will be made publicly available.

Issuance of a license to carry out detailed exploration should carry with it a preference right to production of any minerals discovered. Exercise of this right is described under Principle VI.

B. Competitive Licensing

Where competition exists for detailed exploration licenses, and where the Department of the Interior judges competing exploration plans to be technically sound and the explorers capable of carrying them out, exploration licenses should either be given to all parties or a cooperative exploration program developed in conjunction with the Department of the Interior. Exploration licenses issued in this competitive situation should require all of the elements included in the previous section, except that they should carry no preference right to minerals discovered.

Rationale: The Panel believes that the detailed exploration license is the vehicle whereby the government and the public gain access to detailed resource information. The tracts covered under this license should be large enough to encourage rapid data acquisition (the definition of large will vary with the character of the mineral sought). The general areas where licenses would be available for detailed exploration should have been laid out in the license schedule.

The federal government should establish fixed time intervals within which exploration must be completed. At the end of this fixed time interval the holder or holders of the exploration licenses must apply for a production license, or indicate that they do not wish to apply for one. In either case, the data and its interpretation must be provided to the government. Only those companies having exploration licenses may apply for a production license. The Panel proposes this approach as a means of arriving at a balance between the government's need for information and the need to insure that company-collected information will not be used by other companies unwilling to undertake detailed exploration.

The royalty rate to be charged should be publicly established at the time of the announcement of the availability of exploratory licenses. Procedures for setting this rate are recommended in Principle VI-D.

V. License Impact Statement

Based on the programmatic statement for the region, the Department of the Interior should prepare a license impact statement to be available at least six months prior to production licensing and three months prior to a public hearing. This impact statement will be triggered by the application for a production license. The impact statement should include inputs from the Department of Commerce and other appropriate federal agencies. This statement should be subsidiary to the regional programmatic statement, and be focused on the local area to be licensed. It should address the broad issues raised in the programmatic statement, but concentrate on their immediate or local impacts. It should not repeat material addressing general world or national mineral needs, nor should it address world or national social or economic implications. The one exception is, if the information in the programmatic statement has become obsolete, these specific license impact statements may be used to amend the programmatic statements. The license impact statement should specifically include an assessment of federal management capabilities with regard to these minerals.

Rationale: The data collected by the companies operating under the exploration license, the application for a production license, and the environmental data collected by the government should be used to inform those preparing the impact statement. In cases where there are competing applications for a production license, as covered in the next section, the government will have to make its selection of a licensee in advance of the preparation of the license impact statement. The license impact statement may cover several licenses in the same area, but each separate mining operation should be assessed.

VI. Production License

It should be government policy to license mineral production, as opposed to leasing areas which contains minerals. A major reason for this language change is to distinguish hard minerals regulation from oil and gas regulation.

A. Alternative Licensing Procedures

Production licenses should be given on the basis of work programs submitted by the companies. Such programs should designate the extent of the mining activities to which the company commits itself over a given period of time and in a given area. Additionally, such work programs should also include plans to minimize the negative impact of mining activities on other interests in the area.

1. Non-competitive Licensing

In the absence of competing companies, or where one party has preference rights as described in Principle IV, licenses should be given on a non-competitive basis once the Department of the Interior has judged the proposed work program to be satisfactory.

2. Competitive Licensing in Areas with Known Marine Mining Deposits

Where more than one party desires a license to work a mineral deposit, the license should be given on the basis of competition, based on proposed work programs. Preference should be given to the applicant proposing the most vigorous, and at the same time, the most careful work program. Competition should not be judged on the basis of either bonus bids or royalty rates.

Rationale: A major reason for the Panel's preference for work program competition is to insure rapid development of the resources. Failure to meet the work program, in the absence of compelling reasons, should result in withdrawal of the production license.

In both competitive and non-competitive situations, the government should insure that an adequate work program has been proposed.

B. Relinquishment

At fixed time intervals, substantial percentages of the land covered in the initial license should revert to the federal government. The objective is to encourage potential miners to carry out extensive and early exploration. The major relinquishment should be triggered at the point where the first commercial production begins.

Rationale: The Panel's preference for a relinquishment process is to insure that detailed exploration will occur throughout the tract. It assumes the mining company will choose the richest and most profitable ore for development and relinquish those portions of the tract that fall in other categories. This publicly available information may indicate relinquished ore resources that are attractive to other mining concerns.

C. Information and Data

After issuing a production license, government should have access to all technical data and interpretations held by the licensee. The federal government should have as one of its objectives the accumulation of as complete a data base on subsea minerals as possible. In this connection, government should be encouraged to accelerate its own data collection.

D. Payment

As noted earlier, licenses should not be allocated on the basis of bonus bids nor royalty bids. Rather, royalty rates should be established by the federal government based on the costs and benefits to the licensee. Royalty rates should be low when the risks are high and vice versa. That is, as technology and procedures are developed which bring increasing stability to mining operations, royalty rates should be increased. To assist in setting royalty rates, the Department of the Interior should, at fixed intervals, establish *ad hoc* commissions to assess the adequacy of the royalties being charged mining companies. These commissions should be made up of diverse and financially disinterested persons.

Rationale: The procedures for determining fair return to the government assume that the royalty rate will not change for the duration of the license. The Panel generally used 10-year license periods as its reference point with renewal available. Renewal, however, should be at the royalty rate current at the renewal time. The precise license time period would doubtless need to reflect the character of the operations associated with each specific ore.

E. Liability

The licensee should be fully and absolutely liable for the consequences of his activities. If necessary, the federal government should provide a liability insurance program for claims of an extraordinary kind, i.e., such as that covering nuclear installations.

Rationale: It is the Panel's view that other parties should not suffer uncompensated losses resulting from marine mining, regardless of the existence of negligence or lack of it by the miner. At the same time companies cannot be held liable for activities that may be beyond their capability to obtain insurance. If marine mining is in the public interest, then it seems reasonable that the federal government cover extraordinary claims.

VII. Regulatory Management

Except where prohibited by existing legislation, responsibility for the management and regulation of offshore mining should be concentrated in the Department of the Interior.

The Department should be clearly and publicly responsible for the safety of marine mining operations, and should have sufficient authority and capability to carry out this responsibility. The Department should be assisted by other federal agencies as required.

Rationale: The Panel's recommendation reflects a desire to concentrate responsibility and thereby escape, to the extent possible, the debilitating consequences of having to deal with a multitude of government agencies.

VIII. Standards Specifications

The United States government should support the establishment of an independent standard setting organization, using Det norske Veritas as a model. This organization, located in Norway and serving Norway, Finland, and Sweden, is supported by an extensive and skilled research staff which has distinguished itself by timely study of potential marine problems. The Panel cites this organization as a model for improvement of our domestic standard setting. This standard-setting organization should provide the technological backup for the United States regulation of marine mining.

Rationale: The Panel's recommendation reflects the view that no industry carrying on activities vital to the public interest should be allowed to set its own technical standards for equipment. Adequacy of standards should be determined by specialists without direct economic interest in the activity. This is doubly the case if it is important that those standards be credible with the public. The Panel also believes that such a standard-setting organization can provide a continuing inducement to the industry to improve its technology.

IX. Safety and Environmental Protection

All safety and environmental protection technology used in marine mining operations should meet the "best available" standard.

Rationale: The Panel means by "best available" the best commercially available. Such a standard insures that equipment manufacturers have a ready market for improved safety and environmental technology. Under these circumstances, there is reasonable assurance of a continuing motive for technology improvement over time.

X. Restoration

Each production license shall specify the nature and extent to which a mine must be restored. Assuming that mining licenses will not be issued for more than 10-year periods, each license renewal will specifically review and, if appropriate, modify the terms of the required restoration. The miner should be required to provide an appropriate bond, or contribute to an escrow account sufficient to ensure restoration in the absence of the miner's ability to meet the terms of the license. The special emphasis on continuing review of restoration requirements reflects the present limited data and knowledge in this area.

Rationale: The Panel can find no conclusive evidence of a need for specifying restoration activities for marine mining operations at this time. The above recommendation is included because, while land based mining experience is not directly related to marine mining, the obvious degradation of the terrestial environment where no conditions were attached for restoration, suggests that the need and procedures for restoration be reassessed continuously.

[15] Shepard, Francis P. 1972. *Submarine Geology*, (3rd ed.), New York: Harper & Row.

[16] Fairbridge, R.W. 1966. *Encyclopedia of Oceanography*, New York: Van Nostrand.

[17] National Environmental Policy Act of 1969. Public Law 91-190; 83 Stat. 852:42 U.S. Code 4331 *et seq*.

[18] Allan Hancock Foundation, University of Southern California. 1965. *An Oceanographic and Biological Survey of the Southern California Mainland Shelf*, California Water Quality Control Board Pub. 27, Vol. 1 and Vol. 2 (1971).

[19] Allan Hancock Foundation, University of Southern California. 1971. *Biological and Oceanographical Survey of the Santa Barbara Channel Oil Spill: Volume 1, Biology and Bacteriology*, Isdale Straughan, (ed.).

[20] U.S., Department of Commerce. 1974. *Ocean Instrumentation*, Washington, D.C.: Department of Commerce.

[21] U.S., Department of the Navy. 1961, 1963. *Oceanographic Vessels of the World*, NODC General Series Pub. G-2, 2 Vols, Washington, D.C.: U.S. Government Printing Office.

[22] U.S., Department of Commerce. 1974. *Report of the NOAA Scientific and Technical Committee on Marine Environmental Assessment*, Washington, D.C.: Department of Commerce-Office of Marine Resources.

[23] U.S., Department of Commerce.1972. *Marine Pollution Monitoring: Strategies for a National Program*, Washington, D.C.: Department of Commerce.

[24] Isaacs, John D. 1974. A review of the marine biology of the Southern California offshore region. Talk presented during the Conference on Recommendations for Baseline Research in Southern California relative to Offshore Resource Development, Long Beach, California, December 5-7, 1974. Available from the Southern California Academy of Sciences.

[25] Council on Environmental Quality. 1974. *OCS Oil and Gas - An Environmental Assessment*, 5 Vols., Washington, D.C.: U.S. Government Printing Office.

[26] U.S. Congress. Senate. Committee on Commerce. 1974. Outer Continental Shelf Oil and Gas Development and the Coastal Zone, 93d Cong., 2d session, Washington, D.C.: U.S. Government Printing Office.

[27] Roels, O.A., et al. 1973. The Environmental Impact of Deep-Sea Mining, NOAA Technical Report ERL 290 OD11, Boulder: Department of Commerce.

[28] U.S., Department of the Interior. 1974. Draft Environmental Statement: Proposed Outer Continental Shelf Hard Mineral Mining, Operating and Leasing Regulations, Washington, D.C.: Bureau of Land Management.

CHAPTER FOUR

DEEP-OCEAN MINING

CHARACTERISTICS

The deep seabed beyond the continental shelf is characterized by three general topographies:

- ocean basins, having an average depth of about three to six kilometers;
- seafloor mountains, ridges and fracture lines, representing bold relief above the ocean basins; and
- trench systems, representing bold relief below the ocean basins.

Of these categories, only the ocean basins are of immediate concern to proposed mining in the deep ocean.

The principal deposits under consideration are the ferromanganese nodules, which contain manganese, copper, cobalt and nickel in commercially valuable quantities. It is possible that other deposits, such as phosphorites, ferromanganese crusts and metal-rich deep ocean muds may be exploited in the future. Certainly, much of the capability developed in ferromanganese nodule mining will contribute to making other deep-ocean mining operations economically feasible.

The existence of manganese nodules has been known for some time. The British Challenger Expedition (1870-73) brought back samples of them. However, it was not until the past two decades that the quality and global extent of this material was established. During this same period technology became available that could make possible commercial exploitation of these deposits. But political questions (i.e., the emerging Law of the Sea) complicated the picture from the point of view of private interests which sought to gain and maintain the right to conduct deep-ocean mining on the international seabed.

TECHNOLOGICAL ASSESSMENT

It should be noted during the following review that the problems discussed are not unique to deep-ocean mining. Further this section of the report implies a certain amount of engineering optimism. However, this optimistic attitude should be considered valid only through the developmental phase of ocean mining. The Panel believes that long-term, day-to-day reliable operations for producing ore on a 300-day per year basis are several years away.

During the workshop it was noted that the section on technology of the report is weak as there was little or no discussion of processing. The Panel acknowledges this weakness, and attributes it to the proprietary nature of processing and the lack of details available for publication. The several approaches under consideration by industry at this time are certain to change and offer a relatively broad spectrum. Rather than to publish generalities or make gross estimates, the Panel prefers to omit discussion of processing. In any case during the initial years, processing will be accomplished on land, and the subject thus beyond the scope of this report.

Deep-ocean mining on a production scale is a massive operation. Figure 15 shows a system composed of several elements separated by thousands of miles requiring movement of ore, people and supplies by land, sea and air. Such a system may require as much as 20 years to bring on stream and an investment of between $200-$750 million (1973 dollars).[29]

Figure 16 shows three elements of the mining equipment, and the options that exist to form a mining system. All of these options are under consideration by the United States and foreign firms now active in development of deep-ocean mining systems. Figure 17 shows the typical time-phasing of activity from initiation of research to actual operations.

Commercial mining of mineral resources from the ocean floor involves three basic tasks: ore body location and delineation; mining; and transport of the ore to shore. Each requires a somewhat different technology base and approach. Geological and geophysical exploration and mining operations are similar in many respects, in that some form of surface vessel is required to support each, and both require some form of submerged equipment with suitable navigation of the surface vessel.

Exploration

Published data identifying ferromanganese nodule deposits has been developed for the most part from studies sponsored by federal grants. The data that have received the greatest worldwide distribution are contained in a series of reports sponsored by the National Science Foundation (International Decade of Ocean Exploration) and prepared by Lamont-Doherty Geological Observatory of Columbia University.[30]

Unpublished proprietary data, similar to that in the Lamont-Doherty reports, exist in private industry. Combined, these data provide evidence that valuable mineral deposits exist in several areas of the world oceans. It appears that the

- PRODUCTS
  - Nickel
  - Copper
  - Cobalt
  - Manganese
- PROCESSING PLANT
  - 5000 tons per day
  - 95% waste
- ORE TRANSPORT
  - Barges
  - Ships
- CREW AND SPARE PARTS
  - Air
  - Ship
- MINING SYSTEM
  - Surface Platform
  - Nodule Lift
  - Bottom Miner

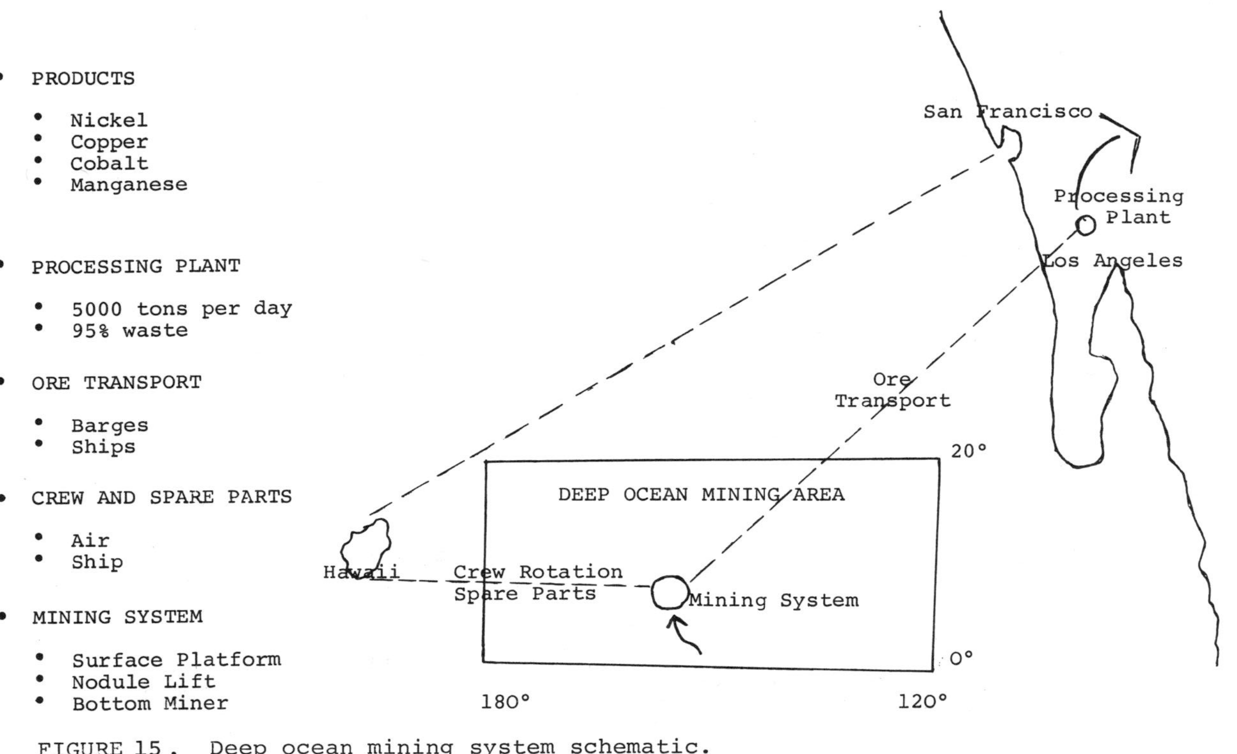

FIGURE 15. Deep ocean mining system schematic.

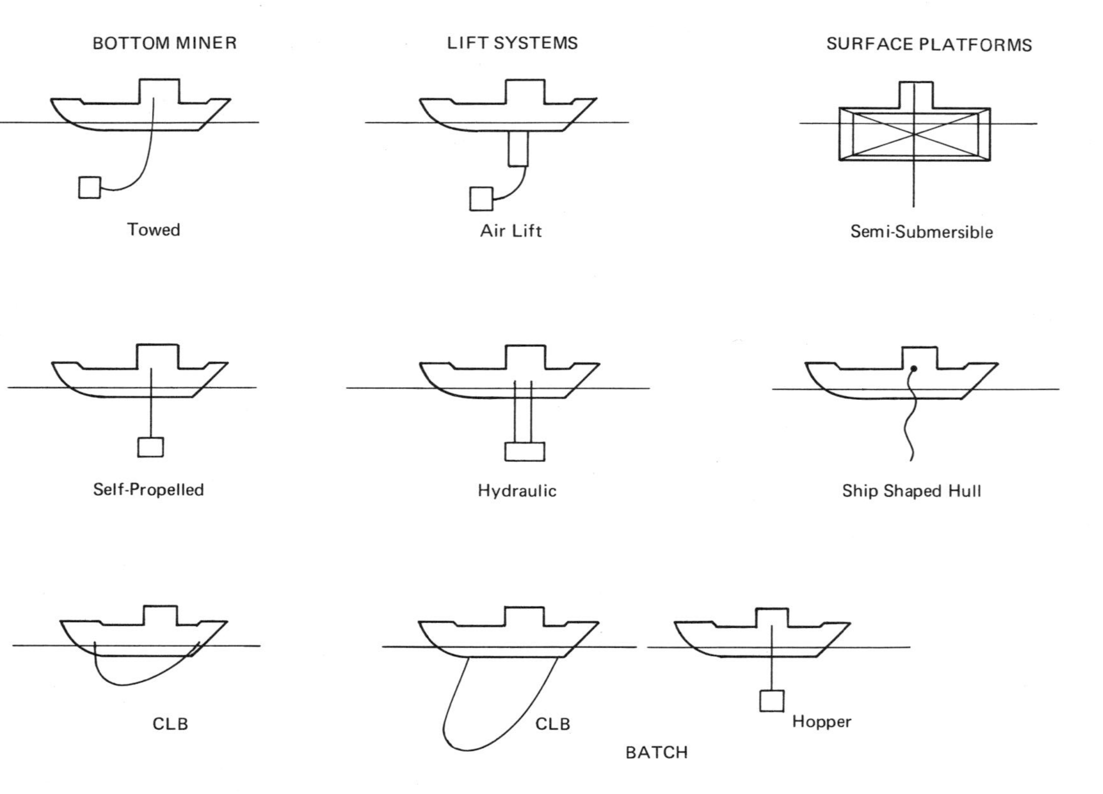

FIGURE 16. Deep-ocean mining approach and options.

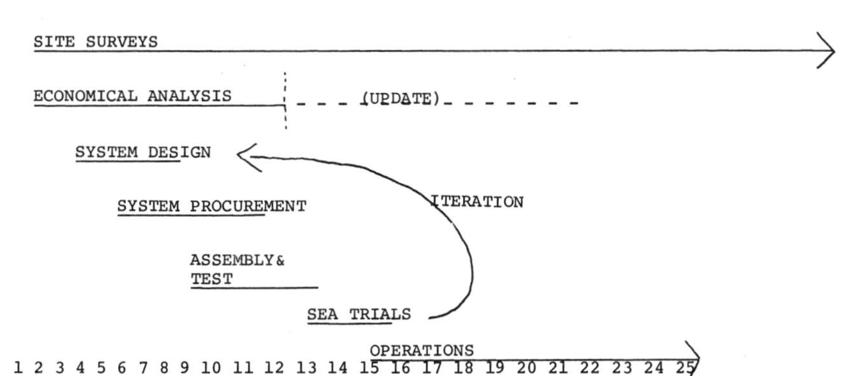

FIGURE 17. Deep-ocean mining, time phasing.

richest deposits are found in the Pacific Ocean basin in water depths of 4300 to 4900 meters (14,000 to 16,000 ft).[31]

The Panel members from industry are in general agreement that the nodules with high content of nickel, copper and cobalt are found in the siliceous (radiolarian) ooze and red clay areas of the Pacific Ocean, which are located in the area of 0°-20°north latitude and 120°- 180° west longitude.[32] It appears that the siliceous ooze host soils produce nodules of approximately twice the nickel and copper content as those found in red clay host soils,[33] but these conclusions remain to be established by actual mining operations.

It should be noted here that the sea conditions in the areas of initial mining interest are generally calm, with sea and swells seldom combining to form wave heights greater than 1.5 meters (5 ft) except in rare extreme storm conditions. However, wave-induced forces are a concern to the miner since these forces directly affect the mining system. For this reason a national weather model is recommended for the areas of mining interest. Since these areas are off the shipping lanes, there is only a limited amount of weather data.

To justify the high initial investment required for ocean mining development, industry must mine the nodules of highest assay first (nickel, cobalt and copper); hence, the areas of initial interest will likely be the siliceous-ooze zones.

Nodule populations are variable and found in a patch-like distribution of irregular geometry. The size of patches sufficiently high in nodule density concentrations and/or assay, vary from several hundred meters to kilometers in maximum dimension. The initial task of ocean mining activities will be exploration on a scale adequate to locate and define deposits of grade and quantity that will support long-term mining projects. These deposits must be located on terrain capable of being effectively traversed by presently contemplated mining devices.

Physical Description of Nodules

Table 6 describes the characteristics of the northeast Pacific Ocean nodules. The nodules in any given common area are fairly consistent in size and mineral content. While these criteria may vary on the order of 25 percent, important ocean mining parameters such as average daily production are predictable enough to allow orderly forecasts of business economics. Data reduction of the Lamont-Doherty Geological Observatory [34] and unpublished reports

Table 6

North East Pacific Ocean

Ferromanganese Nodule Characteristics

Depth of Water Column	3600-5400 meters (12,000-18,000 ft)
Host Soil[1]	Siliceous ooze
Size Range	2.5-23cm (1.0-9.0 in.) greatest dimension
Specific Gravity	1.9-2.4 (drip-dry)
Mn, Ni, Cu, and Co Content[1]	30% Mn, 1.2% Ni, 1.0% Cu, 0.25% Co
Hardness	Range from 3 (calcite to 4 (fluorite) on the mohs scratch scale
Shape	Predominantly oblate spheroid
Moisture Content	Approximately 30%

[1]Values are averaged values from nodules located in siliceous ooze. Red clay host soils have Ni, Cu, and Co values that are approximately one-half that of the siliceous ooze host soils.

have shown that the average density of nodules range from 0.25 to 2.0g/cm^2 (0.5 to 4.0 lbs/sq. ft). A value of 1g/cm^2 (2.0 lb/ft^2) is a suitable engineering value for use in mining rate computations. The size of the nodule patch, will vary from location to location; however, when large nodules such as 23 cm (9.0 in.) in diameter are found, the population is usually small. When small nodules [4 cm (1.5 in. diam)] are found, the population is usually large. Each has the potential yield of the same tonnage when computed on a grams per square centimeter of ocean floor.

Exploration techniques used by government and industry researchers thus far have been adequate to identify grossly the potential value of ferromanganese nodules found in the oceans. Methods used to date include dredges, core samplers, free fall/free surface ascent samplers, prototype test mining equipment, photography, closed circuit television, and acoustic sensors. Each method has been designed to meet the research objectives of the user. The exploration techniques used thus far may not be adequate for the future demands of full-scale mining operations. *In situ*, real-time, nodule analysis is certainly a desirable capability for future mining exploration phases.

Mining Operations

There were over 30 United States patents and a number of foreign patents granted as of 1972 for deep-ocean mining equipment intended for gathering or handling ferromanganese nodules. At least four major United States industrial firms, as well as several foreign groups, are known to be active in development of full-scale ocean mining equipment at this time. Each inventor and industrial firm has actively pursued its unique method for mining nodules and several are certain to function at a level of performance which satisfies the owner.

There are basically two types of dredges, mechanical and hydraulic. The mechanical dredge digs and lifts the material in a container. The hydraulic dredge lifts the material as a slurry in a contained stream of upward-moving liquid. The two systems are competitive in many applications, and the choice of method will depend on the application. Up to now, most deep water dredges have been designed around hydraulic systems with variations in the means of inducing slurry flow in the pipe, in the design of the nodule-gathering device, and in the means of imparting horizontal motion to the system on the seabed.

The continuous bucket line dredge is a mechanical system. A long continuous loop of rope is hung over a platform floating on the surface and the bottom end of the loop is allowed to touch the seafloor; attached to the rope at intervals are ordinary drag buckets. When the loop is caused to rotate, the buckets in their passage will excavate material from the seafloor and carry it to the surface. If the platform on which the system is mounted moves in a direction at right angles to the plane of the loop, then a path equal in width to the length of the platform should be swept across the ocean floor. This simple principle is illustrated in Figure 18.

In a hydraulic system, vertical slurry flow may be induced (1) by moving the slurry directly through pumps installed in the pipeline, or (2) by the injection of air into the line. The latter method lowers the density of the slurry causing an upward flow due to the pressure difference of the fluids inside and outside the pipe. Slurry flow may also be induced by injection of a high velocity stream of fluid into the line in the direction of flow. Other variations are possible, but are not important at this time. Variations in the design of nodule-gathering devices are mainly tied to the method of propulsion of the system. Two basic methods are the towed system, in which the gathering device is towed by the surface platform to which it is attached, and the self-propelled system, in which the gathering device is activated independently of the surface platform. The self-propelled system requires a power source at the seafloor and is necessarily much more complex in design.

This report will not assess these proposed designs since full-scale operations have not been attempted with the equipment. There are several technological problems that must be solved that are common to the success of a given design. These are listed in Table 7.

Mining Equipment and Seabed Interface During and After Ore Removal

The concentration and collection of ferromanganese nodules from the seabed with a minimum related pickup of sediment at high mining rates [5,000 metric tons (5500 tons) per day] is the long-range goal of all deep-ocean mining system designers.

Many methods are proposed to meet this objective; however each must successfully solve the same problems in order to become economically competitive. These problems will be described in the following sections.

1. Nodule Pickup and Seabed Soil (Mud) Removal

The nodule pickup action, of necessity, disturbs the seabed.[35] Suction heads, dredges, buckets, tines, rakes, rollers, and all other pickup mechanisms require some form of scraping, lifting, plucking, pushing, or pulling action on the nodule in order to remove, or transport the soil from the area of mining into the mining machinery as clods of soil, particles in the water column, or soil adhering to

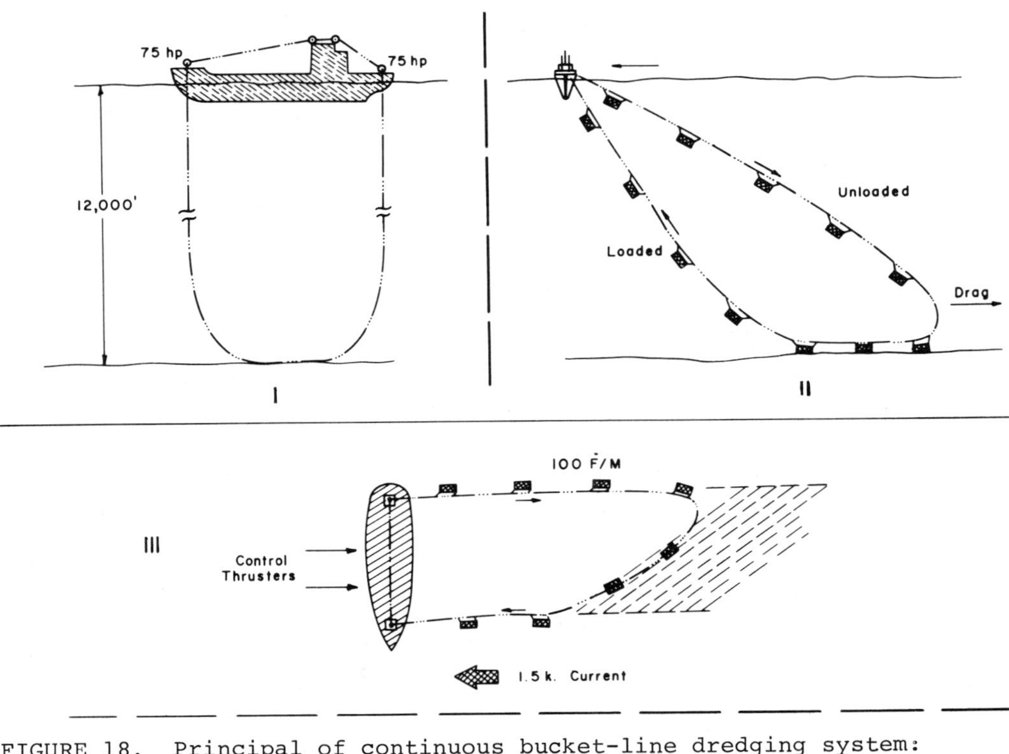

FIGURE 18. Principal of continuous bucket-line dredging system: I. A long, continuous loop is hung over the platform; II. attached to the loop are ordinary drag buckets; III. as the ship moves sideways, a path is swept across the seafloor. (Y. Masuda, M. Cruickshank, and J. Mero, Continuous Bucket Line Dredging at 12,000 Feet)

TABLE 7. Typical mining subsystem.

Element	Risk High	Medium	Low	Remarks
Surface ship			X	State of the art supports
Pipe & pipe handling		X		Fatigue life in sea water is not predictable
Mining equipment		X		Lack of soil mechanics data presents a risk-use of submersible technology solves the remainder
System dynamics	X			Depends upon accurate modeling of complex hydrodynamic forces in order to predict
Nodule lift		X		Depends upon method
Mud removal		X		Techniques are developed
Soil mechanics		X		Verification of theories and literature remain to be tested by experience
Fatigue life of materials	X			Entire engineering discipline is under development
Structures, buoyancy, hydraulics, electrical, electronics, sensors, etc.		X		Use existing deep submergence or offshore petroleum technology

the nodules. Many of the pickup devices perform a stripping action which intentionally removes a swath of soil the width of the mining head and to a depth of several inches. This stripping action is designed to intersect the nodules, which are buried about half way in the soil.

Various methods of soil removal from the nodules will be tested at sea; some candidate techniques are certain to be mechanical shakers, cyclones, water jetting, ultrasonic cleaners, and combinations of each.[36] The system design should accomplish removal at the seafloor. The net result, if successful, will be a sediment plume in and around the mining head, which will settle out in the general area of the seabed that has been mined. The amount of particle scatter will depend upon the design of the mining head, its velocity, Stokes Law and prevailing ocean currents.

An example of the maximum amount of seabed soil that may be disturbed as a result of ocean mining can be computed using the following assumptions:

Nodule Concentration	1 kg/m^2 (2 lbs/ft^2)
Mining Rate (nodules)	4900 metric tons/day (5400 tons/day)
Mining Swath	15 meters (50 ft)
Depth of Swath	10 cm (4 in.)
Velocity	38 cm/sec (1.25 ft/sec.)

On a daily basis this gives a soil (including nodules) pickup of 4.9×10^4 cubic meters (1.7×10^6 cubic ft), along a 30 kilometers (19 miles) swath of 15 meters (50 ft) in width. For a 300-day year, this equates to 14.7×10^6 cubic meters of soil disturbance. This yearly rate is approximately equal to slumping due to turbidity currents at the mouth of the world's large rivers when the river is flooding and wave stirring is at a maximum. For example, the Mississippi River Delta sediment transfer is estimated to be $2x10^8$ cubic meters per year. [37]

The self-propelled mining equipment has an interaction with the seabed to a much greater extent than towed equipment. The self-propelled approach suggests some form of mobility and trafficability that in turn must consider wheels, tracks, or other forms of propulsion. This disturbance of the seabed is probably more uniform with a self-propelled approach than a towed dredge, but is also more likely to disturb a greater volume of seabed when both are operating properly; this is due primarily to swath width and burial depth of the propulsion equipment. Unlike nodule pickup and mud removal, the mobility function will compress the seabed along with stirring up particles.

Using the assumptions previously given, i.e., that of a 4600 metric tons (5000 tons) per day unit traveling 30 km (19 miles) per day, and further assuming a track width on each side of 180 centimeters (72 in.), and a burial depth of 92 centimeters (36 in.), the amount of disturbed soil per day is $1.9x10^5$ cubic meters ($7x10^6$ cubic ft) or $5.7x10^7$ cubic meters ($2.1x10^9$ cubic ft) per year for a 300-day year.

2. Navigation and Obstacle Avoidance

The mining equipment traversing the seabed must either be towed or self-propelled. Some examples of towed systems are the continuous bucket ladder, hydraulic dredge heads, and air lift dredge heads. These systems have the advantage of simplicity and the attendant reliability that simplicity offers. A self-propelled mining system offers the potential of greater efficiency when operating in rich ore (nodule) deposits.

Each approach must consider undersea navigation of the bottom-located mining equipment and geographic navigation of the surface ship. Satellite and inertial guidance navigational aids are available commercially with accuracies adequate for ocean mining surface ships. The problem is compounded because the mining equipment must be controlled with respect to the surface ship.

The navigation function for self-propelled mining equipment also produces a relatively small amount of heat, light and sound in the immediate area of mining. The heat is generated by friction of rotating machinery and lights for viewing through closed circuit television cameras, each with two lamps rated at 400 watts each and producing 126 lumens per watt (10 candlepower/watt at centerbeam). The noise level could approach 100 db/ubar at one yard at 10/kHz with a roll-off of 6 db/octave. These energies will be absorbed by the huge mass of seawater with essentially no effect.

3. Hardware and Technology

Within the limits discussed in this report, hardware and technology may exist to support successful development of a deep ocean manganese nodule mining system. Certain industries have paved the way for deep-ocean mining. For example, offshore petroleum technology development in the 1960-74 period, and the deep submergence programs of manned and unmanned submersibles have provided the ocean mining system designer many "off-the-shelf" solutions to design problems. The offshore oil industry generally provides heavy duty, rugged, highly reliable hardware. The concern of the designer is assured production with long-term durability and reliability. Deep submergence system engineers cannot always utilize these solutions due to an overriding demand for design solutions which minimize weight; thus, large amounts of resources are devoted to qualification testing and use of exotic materials to arrive at the required safety and reliability at minimum weight.

The Panel believes that the nation has the capability to design, fabricate, test and operate deep-ocean mining systems successfully; however, several years of experience must be obtained and only a limited number of firms currently have experienced teams encompassing the necessary professional and technical disciplines for developing prototype deep ocean hardware systems. There are also a limited number of United States contractors with enough experience in deep submergence programs to build a major deep-ocean mining system with their own internal resources.

A review of the status of hardware was made of offshore petroleum and deep submergence industries, using an "off-the-shelf" criterion for availability of components. Table 8 was organized to demonstrate that the state of technology is sound. It is generally agreed by the Panel that while capability exists to design and build deep-ocean mining systems, several areas of technology need improvement. The principal areas requiring improvement fall into three categories crossing several engineering disciplines:

- structural materials;
- component and system reliability; and
- sensor technology.

4. Structural Materials

One of the basic problems that faces an ocean mining system designer is the depth of the water column that separates the surface vessel from the ocean floor. This in turn presents several problems for the structural engineer in designing the strength member required for raising or lowering the mining equipment. The typical solution is usually found to

TABLE 8. Components for deepsea mining.

Buoyancy Material

Material	Cost/Buoyancy $/lb	Remarks
Fluids	0.50	Hydrocarbons such as JP-4, pentane, hexane, etc., offer least cost solution but are not safe or efficient
Spheres		
Steel	3.00–5.00	Become negatively buoyant at ≈ 15,000′
Aluminum	10.00–15.00	Become negatively buoyant at ≈ 18,000′
Glass	7.00–12.00	Will withstand 10,000 psi (22,400′)
Syntactic foam		
37 P.C.F. for 9,000 psi	13.5	Single size glass bubble, low cost resin
34 P.C.F. for 13,500 psi	66.0	Binary bubbles, high strength resin

Structural Materials

Material	Cost–$/lb in Usable Form (December 1973)	Remarks
Steels		
COR 10/A-36	0.40	Used where strength requirements are minimum
4340	1.00	Good choice for pipe material–must quench and temper
HY-80, 100, 140	1.75	U.S. Navy choice for pressure vessels
Marage	2.25	200 ksi strength with no quench required
Aluminum		
6061	1.15	Weldable, aging to T-6 condition. Offers F_{t_u} of 36 ksi
5083	1.25	Best from corrosion standpoint
Titanium		
CP	7.00	Titanium does not corrode in sea water, best choice for long life applications–initial cost is high and difficult to machine and weld
6AL-4V	10.00	
Plastics		
GRP	3.00-5.00	Plastics offer good weight to strength ratios once submerged, lack of marine fouling leads to low maintenance

Hydraulic Components

Item	Performance Available	Remarks
Valves	Off-the-shelf or custom designed	Closed loop, pressure compensated, oil components are S.O.T.A.; salt water components must be custom if not supported by offshore petroleum
Pipe and fittings	Up to 10,000 psig, up to 4″ I.D.	Has not been a problem until exceeding 4″ I.D.
Flex hose	3000 psig up to 2″ I.D.	High pressure, large diameter hose requires development
Motors & pumps	150 S.H.P. can be obtained	Vane, turbine, and radial piston motors; gear, piston, and vane pumps are available off-the-shelf. Oil fluid medium is recommended
Filters	5–10 for oil 150 mesh for sea water	Has not been a problem–suppliers can provide to your requirements

TABLE 8 (Continued)

Sensors

Item	Performance Available	Remarks
Imaging		
CC TV	.75′–100′ in clear water	Low light level cameras are available (silicon intensified target tubes) if used with thallium iodide lamps maximum range results
Navigation		
Acoustics	Several miles (10 KH_z) 100′ (500 KH_z)	Requires narrow bandwidth channels at low frequency
Obstacle avoidance		
Pulse sensors	¼°, 150′ range (650 KH_z)	Requires highly skilled operators in order to integrate typical displays in real time
CTFM	2°, 1500 yards (80 KH_z)	

Electrical Power Equipment

Item	Performance Available	Remarks
Batteries (secondary)	Any reasonable requirements can be met	Several types exist; lead-acid, nickel-iron, nickel-cadmium, silver-zinc, silver-cadmium. Lead-acid is lowest cost approach
Switch gear	Low reliability	Arc across contacts change with pressure-environmental testing recommended
Connectors and penetrators	Man rated quality	37 and 55 pin available with mil-spec standards, all leak paths double sealed
Cable	Any reasonable requirements can be met	Coaxial, twisted shielded pairs, 1/0 power conductor and be made in combinations up to 3″ O.D. with armoring to support 16,000′ length in water

Electronics

State-of-the-art for ocean mining is the same as other industries
- Electronic equipment is located within pressure vessels
  - Access for test and troubleshooting becomes limited
  - Heat transfer can become a problem
- Certain equipment can be housed in pressure compensated containers, if they can operate in oil

Command and control data link is the most complex electronic design task in an ocean mining system

be the use of pipe, if large factors of safety are required, and wire rope, if operating depths are shallow. The choice of the usage of pipe presents problems such as the presence of localized high stresses at the pipe connections, which may be threaded joints. For economical operation, with reasonable trip times, the pipe handling system can become automated or semi-automated. This is one risk area where existing offshore petroleum techniques can be applied.

Because of the implicit complexity, high strength and low density structural materials are in demand. When cost is considered, selection of a pipe string material usually ends with a steel alloy. The engineer can appreciate the need for rigorous trade-off studies supporting proper selection of the pipe string material. The steel producers in this country can currently supply a pipe string at the reasonable costs [$4.40/kg ($2.0/lb)], and in cross sections large enough to permit mining equipment weights in excess of 230,000 kg (500,000 lbs) to be raised and lowered to and from depths of 6000 meters (20,000 ft), with adequate provision for dynamic load amplification and safety factors.

Materials selected for the mining equipment used in the seafloor must conform to a different set of criteria from that of the wire rope or pipe briefly discussed above. Corrosion-resistant materials, or those with a low maintenance cost will probably be preferred for this equipment. The use of dissimilar metals, which can introduce corrosion, must be avoided in designing the equipment, or at least be recognized and counteracted with preventative repair/replacement maintenance procedures.

Three areas of materials technology require a considerable amount of improvement for marine mining applications:

1. fatigue life of materials in seawater;

2. fracture mechanics of materials exposed to seawater and other corrosive media at high stress levels; and

3. residual stresses due to welding.

At first these might appear to be similar problems, but examined carefully they demand different design solutions. This not only affects ocean mining engineers, but all product and research engineers designing underwater equipment as well.

The first area of technological weakness--fatigue life of materials exposed to seawater--requires a testing program that provides published materials characterization data for structural designers. This testing program should evaluate the following parameters: material; mean stress level; stress ratio; cycles to failure for various stress ratios; values in air compared to values in seawater; weldments; forgings; heat treatment; grain direction; and fracture toughness. The permutations available from such a mix of parameters can easily become a test program requiring in excess of 10^5 samples; however, if the candidate materials are limited to the few in general use for structural applications today, the test program becomes one of manageable proportions. The following sample structural materials list is suggested as reasonable.

- Steel - A-36 and/or A-242,

- 4330 (100-150 KSi F_{t_y}),

- 4340 (100-150 KSi F_{t_y}),

- T-1 (90-100 KSi F_{t_Y}).
- Titanium - 6AL-4V
- Aluminum - 5083, 6061

The second technology weakness--fracture mechanics of materials exposed to seawater--requires emphasis at the research level. Fatigue life and fracture mechanics of materials in seawater do interact. The cause of structural failure of a material under tensile stress is understood to be growth of a flaw of subcritical size that causes failure of the structural member in the presence of continued stress. The entire engineering discipline of fracture mechanics has just recently been receiving a level of research effort equal to its importance; however, the development of data for the ocean environment has been given low priority.

The final weak area, residual stresses and corrosion resistance modification due to welding, also requires emphasis at the research level. A generally acceptable method does not exist for predicting or measuring residual stresses due to welding techniques, even in the simple joints of butt, lap, or tee configurations, since complex and indeterminate structures are common weldments in ocean mining structures.

Looking at the first example discussed in this section, selection of pipe string materials for ocean mining systems, with relation to fatigue life and fracture mechanics data, one can see the limited options available to a cost-conscious engineering decision-maker. Without the availability of fatigue life and fracture mechanics threshold data for accept/reject criteria, the structural engineer must make selections for marine materials without benefit of a sound scientific foundation, resulting in heavy and costlier solutions. Lack of data in weldments further compounds the problem.

5. Component and System Reliability

Experience in operating complex machinery in the deep ocean is limited, and successful long-term operations are few in number when the task must consider simultaneous operation of a surface ship, suspension strength member, and some form of towed or self-propelled mining head. The restricted availability of proven hardware at the component level can only add to the risk of deep-ocean equipment operations. Deep-ocean mining when compared to that of such deep-ocean activities such as submersible rescue or search missions, contains the need for a unique reliability criterion; that of long duty cycles and continuous daily operation without the need for raising the mining head from the seabed to the surface.

One approach to solving the system design problem and providing the high reliability necessary for deep-ocean mining equipment has been to use several techniques, each having been proven successful in systems other than mining.

Such components as electrical, electronic, and hydraulic parts are protected from saltwater and hydrostatic pressure if housed within a pressure-proof container which provides a dry gas/component interface. If a pressure-compensated system is used, the electrical and moving parts are exposed to a low pressure over ambient by a compensated, closed loop, inert fluid system. Typically, electronic components such as printed circuit boards, solid-state relays, integrated circuits, switches and terminal boards are located within pressure vessels. For deep-ocean applications, the pressure vessel is usually spherical in configuration, flanged at the equator for access to the equipment. The flanged joint and wire penetrations will usually be sealed with "O" rings. The net result is containment of the equipment within an atmospheric pressure environment. This is usually the environment of the original component design; thus, in the pressure vessel we have simulated the original intent and maintained the initially achieved reliability.

Pressure compensation physically isolates components from seawater but exposes them to ambient pressure plus a small over pressure [215-354 g/cm^2 (3-5 lbs/in^2)] within the compensated loop. The overpressure is constant, independent of depth, and usually mechanized by bladders located below the compensated equipment (head), or spring loaded containers located at the same depth or above the equipment to be compensated. Thus the need for high pressure seals is eliminated. Such a design provides for a fluid (usually oil) medium which bathes the equipment in an environment conducive to high reliability.

Conservative design often results from a prediction of poor reliability performance of a material or component. Several reasons force conservative design. Structurally, in order to provide for low-stress levels, the hydraulic designer consistently uses a "fail-safe" approach to valve selection, and electrical power designers fuse circuits to the point where one questions the reliability of the fuses. Conservatism to a level consistent with cost is good engineering practice. The designer can prevent an "overkill" approach if allowed a rigorous simulation or test program that increases his confidence in the prototype equipment and therefore the equipment to be used. Testing to simulate the deep ocean environment can become costly if hydrostatic pressure, ambient temperature, and data recording equipment are necessary. However, the alternative of testing at sea is more costly and time-consuming. To some extent, the future is limited by the lack of components specifically designed for use in ocean mining systems.

6. Sensors

Operation of mining equipment in the deep-ocean environment imposes the functional requirements which employ various sensors. Underwater navigation is necessary for positioning the mining equipment with respect to a surface support ship, ocean bottom landmarks, or geographical coordinates as required by the mining operation. An imaging capability is needed for high-resolution examination of the ocean bottom, observation of mining machinery operation, and detection of obstacles in the case of mobile machinery. In addition, some operations will also require environmental sensors.

The underwater navigation function generally requires data with low information content, i.e., underwater range and bearing at relatively low update rates. This implies systems sensors that transfer data in narrow bandwidth channels amenable to low frequencies. These characteristics are associated with acoustic systems that operate at frequencies in the kilo-hertz range. The navigation function is performed exclusively by acoustic devices, not only because acoustic frequencies are suitable for this function, but also because the ranges achievable in the ocean medium are orders of magnitude greater at the lower frequencies. Achievable ranges vary from several kilometers for low frequency (less than 10 kHz) range measuring devices used for navigation, to less than 30 meters (100 feet) for relatively high frequency (500 Hz), relatively high resolution(2^{0} beam width), pseudo-imaging devices used for landmark recognition and obstacle avoidance.

The imaging function requires data with high information content; for example, images containing a large number of points at different levels of intensity. This implies a wide bandwidth, high frequency system such as those of optical systems which operate at frequencies in the megahertz range. Imaging functions are performed primarily by optical systems, except for the overlap in the observation of large, far-field objects where the limited resolution of acoustic systems is offset by the limited range of optical systems (which is less than 30 meters (100 ft) for clear water, degrading rapidly if mud and silt are present).

Specific requirements, as well as methods and equipment available to satisfy these requirements, are important for both navigation and sensor applications. The navigation function invariably requires the location of one point or object with respect to a local reference. If the geographical location of the local reference is known from a prior survey, then the geographical location of the unknown can be determined. Various functional schemes are available including transmission of pulse and receipt of pulse returned by a transponder; and/or receipt of pulse from a beacon and comparison with time references. Range is obtained by measuring pulse travel time, while bearing is generally obtained by triangulation using ranging data to two or more points. Devices are available that measure bearing directly by use of a dipole receiving array and measurement of the phase difference at the receivers. Current state-of-the-art equipment provides the following capabilities:

. Range:	Several kilometers with the use of transponders.
. Range Accuracy:	30-300 centimeters (1-10 ft), depending on range.
. Direct Bearing Measurements	5 degrees

Acoustic imaging, in contrast to navigation, strains the capability of state-of-the-art equipment. The high resolution required for imaging demands narrow beams, which require high frequencies, which in turn limit the range. In addition, minimum pulse length limits the achievable range resolution. For example, the absorption of 100 Hz is only $1.09x10^{-4}$ db/km (10^{-4} db/k yd), whereas at 10 kHz the absorption is approximately $6.52x10^{-1}$ db/km ($6x10^{-1}$ db/k yd). Narrow beam transducers generally form a fan-shaped beam which must be received on an orthogonally-placed hydrophone array to achieve a real resolution. State-of-the-art pulse transducers can generate a fan beam one-quarter degree wide at a frequency of 80 kHz with a range from 430 to 1300 meters (500 to 1500 yds), depending on target strength. Continuous transmission frequency modulated (CTFM) systems are generally more complex and costly than pulse systems. If the object to be observed is located on or near the bottom, the problem is further complicated by the need to distinguish returns from the object from bottom reverberations.

True high-resolution imaging can only be performed optically. State-of-the-art television systems provide a resolution of 0.1 degree. The major obstacle in using television is the limited range caused by the scattering and absorption of light. It is therefore advisable to use a light source that suffers the least absorption. The selection of a light source must also take into account the frequency sensitivity of the camera to be used. The most sensitive low light cameras available use Silicon Intensified Target (SIT) tubes. The optimum frequency light that minimizes absorption and maximizes camera sensitivity is produced by thallium iodide lamps at a wavelength of approximately 4500 angstroms. The combination of SIT cameras and thallium iodide lamps operating in clear water, such as is found in the undisturbed deep-ocean, provide viewing ranges of up to 56 meters (185 ft).

There is need for sensor improvements in the following areas:

- Long Range Viewing

Viewing of the subsea terrain is currently limited by the range capability of high resolution optical sensors or the resolution of long range acoustic sensors. The range of optical sensors is limited by the scattering and absorption of light in the medium. Practical methods of overcoming this limitation by polarization, gated receiver and light sources, the use of coherent light, or other means, are needed. Resolution of acoustic sensors could be improved by use of high frequencies and provision of the additional power required, or by other techniques including non-linear acoustics and acoustic holography now in early development.

- In-Situ Analysis of Minerals

Manganese nodules vary considerably in the content of the metals of primary interest, i.e., copper, nickel and cobalt. It would be desirable to obtain an assay without the time-consuming operation of bringing the nodules to the surface. An analysis system to rapidly perform assays on the ocean bottom is desirable.

Additional Systematic Factors of Environmental Interest

A mining rate of 5000 metric tons (5500 tons) per day appears to be a reasonable model for analysis of future ocean mining systems in terms of realistic economies and in the determination of loads placed upon the environment by ocean miners. This rate is equal to the handling of 1.5×10^6 metric tons (1.64×10^6 tons) of raw material (nodules) per 300-day year. The material must be moved from the sea-floor, up the nodule transfer conduit, stowed aboard ship, ultimately transported to shore, offloaded from a barge (or the mining ship) and land-transported to a shore-based processing plant. Since the nodules as recovered from the ocean floor are 30 percent entrained water (by weight), there may be reduced rates for handling as drying occurs, this lower limit being 4.5×10^5 metric tons (4.9×10^5 tons) per year of the nodules that are completely dried. The amount of water transported from the ocean floor for a hydraulic lift system can be computed using the rule of thumb of a maximum of 20 percent solids concentration or four times the nodule tonnage rate which is 6×10^6 metric tons (6.5×10^6 tons) of seawater per 300-day year. Excess water will probably be discharged at depths of 300 to 900 meters (1000 to 3000 ft) below the surface.

ENVIRONMENTAL PROTECTION AND SAFETY

There is no doubt that environmental considerations and arguments, with or without sound technical basis, will be used in international legal, political and economic deliberations concerning exploitation of the mineral resources of the seafloor, as has already been the case in the United Nations Seabed Committee.

Several mining tests have already been completed and many more are in preparation. The prospect of imminent and extensive deep-ocean mining requires serious consideration of the environmental impact of this activity, since it could affect the benthic and pelagic environments. It is essential that the environmental implications of manganese-nodule mining from the deep-ocean floor be thoroughly understood, evaluated and documented before such mining is attempted on a large scale.

The proposed mining of manganese nodules from the deep-ocean floor has triggered a unique collaboration in the United States among the government, mining industry, academic institutions, and public interest groups to determine the environmental impact of the proposed mining operations before their start. This is in great contrast to other important industrial developments, where environmental concerns have usually arisen after damage to the environment--sometimes serious-- has taken place. By taking preventive action, it should be possible to reduce greatly or avert completely environmental hazards due to mining operations. Collaboration between government, industry and academia to ensure safe deep-ocean mining methods could also lead to the development of mining techniques which would not have degrading environmental effects.

The emphasis of this discussion is the impact of manganese-nodule mining on the marine environment. The metallurgical operations to extract valuable metals such as copper, nickel and cobalt from manganese nodules, at sea, should be roughly comparable in their environmental effects to land-based operations of a similar nature. However, if the ore processing takes place at sea, special precautions for the discharge of waste materials will be necessary. Since secondary land use (including land-based processing plants and tailings disposal sites), and social and demographic patterns affected by marine mining or ore processing, are not exclusive problems of deep-ocean mining, these considerations are outside the scope of this report. Similarly, the environmental impacts of alternative means of obtaining metal ores and the environmental analysis of the utilization of minerals obtained from the marine environment are not considered here.

Deposits of manganese nodules of current commercial interest lie mainly on top of the sediments covering the ocean bottom; therefore, no deep penetration of the sediments will be required to retrieve them. Manganese nodules are rare in areas where there is rapid sedimentation--e.g., on those parts of the seafloor underlying areas of high biological productivity in the water column, which give rise to rapid sedimentation of biogenic oozes. The areas to be mined will be limited, therefore, by the distribution of manganese nodules on the ocean floor and by technical and economic factors governing their retrieval from the depths. Our discussion, therefore, considers only relatively flat, sediment-covered parts of the ocean floor with a high density of manganese nodules on, or very close to, the surface of the sediment.

Environmental Impacts of Deep-Ocean Mining

Mining Methods

In a mining operation, the manganese nodules are collected from the ocean floor, usually from great depths, and transported through the water column to a surface vessel. The collection of manganese nodules will result in the removal and redistribution of sediments and benthic organisms on the ocean floor. In all mining operations, it is likely that there will be considerable resuspension of sedimentary materials in the near-bottom waters. All of the different techniques under consideration for nodule mining will try to avoid, as much as possible, the retrieval of sediments with the nodules. The continuous bucket line system tested in the Pacific in 1971 and 1972, used buckets of 40 cm (16 in.) deep with a maximum penetration into the sediment of about 20 cm (8 in.); however, penetration probably will be much less in practice.[38] Other systems propose to utilize bottom-gathering devices connected by hydraulic or airlift pumping systems to transport the nodules to the surface through a pipeline.[39] All of these devices have components that make contact with the ocean floor in separating the nodules from the surrounding sediment. First separation is achieved by a chute with water jets, heavy spring-rake tines, a radial tooth roller, harrow blades and water jets, or spaced comb teeth. Many of the concepts employ adjustable collecting elements so that changes can be made during the mining operation to accomodate variations in the nodule deposit and sediment characteristics. A second important feature of all of the collecting devices is a controlled digging depth into the ocean floor, since interest is usually centered within the upper few inches of the sediment.

A quantitative estimate of sediment resuspension by towed suction dredge heads and by self-propelled mining equipment, operating on the seafloor, are given in the Technology section of this chapter.

Effects of Mining on the Seafloor and Near-Bottom Water Mass

It is in the interest of a mining operation to separate the nodules from the sediment to the greatest extent possible on the ocean floor, and to disturb the sediment as little as possible, compatible with efficient collection of the nodules. However, it is equally obvious that significant disturbance of the sediment and the sessile benthic organisms, that cannot escape the oncoming dredge, will occur. A cloud of sediment will undoubtedly be disturbed in the near-bottom water layers. The distribution and resedimentation of the disturbed particles will obviously be governed by their density and other sedimentation characteristics as well as by the near-bottom currents. This resuspension of sedimentary materials will influence the near-bottom water mass, and certain areas of the ocean floor from which sediments have been removed, and other areas where redeposition of the sediment will occur.

The near-bottom water mass may retain in solution certain compounds leached from the sediment or from the interstitial water. For instance, in manganese nodule areas, it is conceivable that the trace-metal content of the near-bottom water could be increased by the resuspension of sediment. This enrichment of the near-bottom water in certain compounds may have an effect on organisms living in the deep ocean near the seafloor. On the whole, important effects seem unlikely both in view of the relatively low density of the near-bottom prowlers and the fact that the sedimentary material was previously settled on the seafloor as a result of natural sedimentation processes. It has been argued that the redistribution of sediment on the ocean floor resulting from natural phenomena exceeds by many orders of magnitude on a world-wide scale, any disturbance caused by all the dredges ever likely to be utilized in deep-ocean mining.[40] It remains equally clear, however, that local disturbance of sediment may have a certain impact on the deep-ocean fauna and flora. This is particularly the case for sessile animals which may have a very slow reproductive cycle. On the other hand, it is unlikely that any mining operation will cover 100% of a given area of the seafloor. Seafloor bands of adequate width with full consideration given to possible sediment drift, could thus be left undisturbed in a mined area to enable the re-establishment of deep-ocean fauna and flora in those areas where the dredge heads have destroyed it. This process of recolonization might be quite rapid on a geological timescale. It is believed that the biomass of the sessile fauna on the deep-ocean floor is generally very low, particularly in manganese-nodule areas and, therefore, the quantitative impact of deep-ocean mining on the total marine flora and fauna of the oceans should be quite small.

Another possible result of the disturbance of the sediments and their resuspension in the water column is the transplantation of spores or other dormant or live forms of micro-organisms from one area, where they rest in the sediment, to another, transported by water currents in the overlying water masses after resuspension from the dredged sediments. Some of these species are dormant in the sediments but may revive when discharged into other environments. Initial observations on some dormant organisms occurring in deep-ocean sediments have been described.[41]

Effect of Mining on the Water Column

The effect of sediment and near-bottom water discharged at the surface has been measured or forecast.[42] To date, there is no information concerning the rate of sedimentation of discharged particulate matter. There is, however, information concerning the influence of deep-ocean sediment on the productivity of water in the euphotic zone. The influence of dissolved nutrients from interstitial water in nodules, or from near-bottom water, on the chemical composition of the overlying water column can be calculated from the rate of mixing and the fate of near-bottom water at the time of discharge, as well as by the salinity and temperature of the receiving water mass. The volumes of near-bottom water required to lift the nodules from the bottom to the surface in hydraulic or air lift-pump mining systems are given in the Technology section of this chapter.

From the incomplete results of published work to date, it appears that the environmental effect of mining operations; the vertical transport of manganese nodules, sediment, and near-bottom water to the surface; and sediment discharge of the surface or at intermediate levels in the water column, may be small.[43] It should be stressed, however, that the environmental impacts of the resuspension and resedimentation of stirred up and discharged sediment have not been measured at this time.

The processing and extractive metallurgy of manganese nodules at sea, and the discharge of waste materials resulting from this processing, could be dangerous to the environment unless adequate precautions are taken. However, most major concerns involved in the development of manganese nodules have determined that, at least for first-generation plants, economical processing can only be accomplished ashore.[44] The principal reasons for this are that the transportation costs of materials for processing will be equal to, or greater than, the cost of transport of nodules, and problems of waste disposal and environmental protection will be much greater at sea than on land. However, should all processing take place at sea, the care taken in waste disposal resulting from metallurgical processes should be, at the very least, equal to that of land-based operations of a similar nature.

REGULATIONS AND LEASING

Introduction

The philosophy presented in the introduction of the section on the regulations for the outer continental shelf is carried forward in this section. Although those engaged in developing this new resource opportunity are not currently hampered by the Outer Continental Shelf Lands Act, there is concern that it may "creep" from the shelf to the deep-ocean floor through expediency. Rather than struggle within these constraints, the Regulations subpanel has modified the proposed outer continental shelf approach to recognize the fundamental difference in orebody rights, those of the United States government in the case of the outer continental shelf, and those accruing to the miner in the deep-ocean beyond national jurisdiction. Since it is not known at this time how regulation of the deep seabed will be implemented, these suggested regulations are intended to be useful to both the Executive Branch and the Congress.

Beyond 200 meters (650 ft) water depth, the question of who owns the minerals is unclear. That question is presently under negotiation within the international community and under discussion in the United States. Some observers expect that United States companies will go into the deep-ocean and mine minerals in advance of a formal international agreement. Congress is considering action which, in effect, authorizes and controls such activity. It is the view of the Panel that passage of the Deep Seabed Hard Minerals Act is quite likely. It is assumed that should such mining take place, United States companies will be regulated by the United States Government. Under these circumstances, the United States will assert its right to regulate because the miner is a United States natural or juridical person.

The Panel's approach to regulation has been to trade the financial advantage to the government of bonus or royalty bid allocation by sealed competitive bid, for a system which will provide early information and informed management. There is responsibility for the government to perform its statutory tasks in an informed fashion; therefore, there must be an adequate flow of data from the ocean miner to the responsible government agency. It is doubtful that enough quantitative data on deep seabeds exists, at present, to permit effective "resource management" by the government.

Performance standards are required which provide industry with incentives to improve its technology. Improved technology will result in appropriate rewards to both government and industry in the form of increased economic return and enhanced protection of the environment.

The principal objective behind these proposed regulations is the creation of a system of procedures that will make management of these resources responsive to a broad set of public and economic concerns. To accomplish this continuous updating of all kinds of information, continuous improvement of the technologies being utilized, and continuous checking on both the performance of the companies and the performance of the government's agency is required. We believe that at an early stage in the development of an industry, the only real hope is in the creation of a flexible and imaginative set of procedures for managing these resources.

Regulatory Principles Covering Hard Minerals Mining in the Deep Ocean

Prospecting (No Government Permit Required)

Prospecting, using such methods as magnetic, gravimetric, and seismic surveys, as well as bottom sampling and shallow coring, should be available to any United States citizen or company without prior approval. The principle of freedom of the seas applies here.

Environmental Impact Statement (Refer to Pages 63-66)

Exploration-Production License

It should be the policy of the federal government to license mineral production and exploration in the deep-ocean, as opposed to leasing land containing minerals. As in the case of the outer continental shelf (pages 75-76) there should be a regional programmatic environmental impact statement prior to issuing an exploration-production license in any region. The region may be identified by oceans, areas of oceans, geological structure or kinds of orebody. There should also be a specific detailed license environmental impact statement at the point that a company converts from exploration to production (pages 78-79).

Alternative Licensing Procedures

Exploration-production licenses should be given for an initial period of 20 years, renewable for an additional 20, on the basis of work programs submitted by the companies. Such programs should designate the extent of the mining activities to which the company commits itself over a given period of time and in a given area. Additionally, such work programs should also include an effort to minimize the impact of production activities on other interests in the area.

Information and Data

After issuing an exploration-production license, the federal government should have access to all technical and environmental data on the license area held by the licensee. Technical data shall be treated as proprietary by the federal government and by environmental monitoring entities or consultants, at least until relinquishment, abandonment, or termination of the license. Environmental data should be released to the public as rapidly as practicable, with due consideration for the proprietary nature of the technology.

Data to be held as proprietary should be those which are directly related to technical and engineering development and the geological and geophysical information characterizing the resource area. Information relevant to the environment, or to alternate uses of the area, should also be made available. Responsibility for determining data to be released should rest with the Department of the Interior.

The licensee should be obligated to provide significant information on environmental or multiple-use concerns associated with his operations as a part of the notification procedure. Additionally, any significant changes in the type, quantity, or quality of production operations from those described in the license application should also be included in the notification that production operations will be undertaken.

Payment

Licenses should not be allocated on the basis of bonus or royalty bids. Rather, royalty rates should be established by the federal government based on the costs and benefits to the licensee. Royalty rates should be low when the risks are high and vice versa. That is, as technology and procedures are developed that bring increasing stability to operations, royalty rates should be increased. To assist in setting royalty rates, Interior should, at fixed intervals, establish *ad hoc* commissions, to assess the adequacy of the royalties being charged mining companies. These commissions should be made up of diverse, and financially-disinterested, persons. Particular attention should be paid to the establishment of the *ad hoc* commission.

Relinquishment

At fixed time intervals, substantial percentages of the resource rights covered by the initial license should be relinquished. This would encourage extensive and early exploration and provide maximum information. The major relinquishment should be triggered at the point where the first commercial production begins. A 50 percent relinquishment is considered appropriate for early licenses.

Regulatory Management

Except where prohibited by existing legislation, responsibility for the management and regulation of offshore mining should be concentrated in the Department of the Interior. Interior should be clearly and publicly responsible for the safety of marine mining operations and have sufficient authority and capability to carry out this responsibility.

Standards Specifications

The United States Government should support the establishment of an independent standard-setting organization using Det norske Veritas as a model. This organization should provide the technological support for United States regulation of marine mining.

Safety and Environmental Protection

Safety and environmental protection technology used in marine mining should be the "best available" commercial standard, with adequate provisions made for observation and monitoring the various environmental effects of mining previously outlined.

[29] Meiser, H.J. and E. Miller. 1973. Manganese Nodules: A Further Resource to Meet Mineral Requirements?. Papers on the Origin and Distribution of Manganese Nodules in the Pacific and Prospects for Exploration, Maury Morgenstein, (ed.), pp. 23-25.

[30] Horn, D.R., et al. 1972. Ferromanganese Deposits of the North Pacific Ocean, Palisades, New York: Lamont-Doherty Geological Observatory.

[31] Dietz, R.S. 1955. Manganese Deposits on the Northeast Pacific Seafloor. Cal. Jour. Mines and Geol., Vol. 51, pp. 209-220.

[32] Horn, D.R., et al. 1972. Ferromanganese Deposits of the North Pacific Ocean, Palisades, New York: Lamont-Doherty Geological Observatory.

[33] Buser, W. 1959. The Nature of the Iron and Manganese Compounds in Manganese Nodules. International Oceanography Congress Preprints, p. 962.

[34] Horn, D.R., et al. 1972. Ferromanganese Deposits of the North Pacific Ocean, Palisades, New York: Lamont-Doherty Geological Observatory.

[35] Roels, O.A., et al. 1973. Environmental Impact of Deep-Sea Mining, NOAA Technical Report ERL 290-OD11, Boulder: Department of Commerce.

[36] Welling, C.G. 1972. Some Environmental Factors Associated With Deep Ocean Mining. Preprints of the 8th Annual Marine Technology Society Meeting, Washington, D.C.

[37] Shepard, F.P. 1972. Submarine Geology, (3rd ed.), New York: Harper & Row.

[38] Masuda, Y., *et al.* 1971. Continuous Bucket-Line Dredging at 12,000 Feet. *Offshore Technology Conference Preprints*, Vol. II, pp. 837-858.

[39] Garland, C. and R. Hagerty. 1972. Environmental Planning Considerations for Deep Ocean Mining. *Preprints of the 8th Annual Marine Technology Society Meeting*, Washington, D.C.

[40] Garland, C. and R. Hagerty. 1972. Environmental Planning Considerations for Deep Ocean Mining. *Proceedings of the 8th Annual Marine Technology Society Meeting*, Washington, D.C.

[41] Malone, T.C., *et al.* 1973. The Possible Occurrence of Photosynthetic Microorganisms in Deep Sea Sediments of the North Atlantic. *Jour. of Phycology*, Vol. 9, pp. 482-488.

[42] Amos, A.F., *et al.* 1972. Deep-Ocean Mining: Some Effects of Surface Discharged Deep Water. *Papers from a Conference on Ferromanganese Deposits on the Ocean Floor*, D.R. Horn (ed.), Washington, D.C.: National Science Foundation.

[43] Roels, O.A., *et al.* 1973. *Environmental Impact of Deep-Sea Mining*, NOAA Technical Report ERL 290-OD11, Boulder: Department of Commerce.

[44] Cardwell, P.H. 1973. Extractive Metallurgy of Ocean Nodules. Paper presented during the Mining Convention/Environmental Show of the American Mining Congress, 9-12 September, 1973, Denver, Colorado.

CHAPTER FIVE

EDUCATIONAL CONSIDERATIONS: MANPOWER FOR OCEAN MINING

INTRODUCTION AND SCOPE

The Panel initially considered several different aspects of education, including direct public information and the public's view of marine mining as projected through the mass news media--television, press and radio. In the end, the Panel chose to center its attention on higher education as a manpower service for ocean mining, including the attendant activities of marine minerals exploration and mining regulation and monitoring. The Panel believes that providing the specially trained manpower necessary to the future development and wise administration of ocean mining should be a national concern. Moreover the Panel judges that education *per se* does not make the distinction between mining in deep water versus continental shelf; thus, this part of the study cuts across the at-sea operational boundary.

As an embryonic industry, marine mining has fostered only the beginnings of specific institutions, educational materials, and tailored curricula. The engineers and technicians currently engaged in frontier development of marine mining have been trained in other disciplines, and have simply extended their earlier training and experience to meet marine mining demands. This is not unlike the earlier efforts of land-oriented biologists, geologists, and physical scientists who, earlier, entered oceanography. Nevertheless, the demands of marine mining a decade hence, and possibly much sooner, do require that the full educational spectrum be considered. This should include public education through the executive tier of corporate, congressional and agency organizations, and through the operational levels for the training of marine mining specialists in the professional category, technicians, operators and supervisory managers. It is a large order.

As with any new industry during its early growth, marine mining must draw upon existing educational processes and institutions. In this instance, the related activities are ocean science and engineering, mining, offshore petroleum production, minerals exploration, environmental and resource sciences, economics, marine affairs, and admiralty and international law. In assessing the current state of the present educational spectrum related to these activities, more was drawn from the experience and knowledge of the Panel than from documented study. The Panel examined the needs for informing the public and for educating future marine mining manpower.

PUBLIC UNDERSTANDING

In order for the public to support ocean mining activities, it needs to be informed with accuracy and credibility. To assess the present state of public understanding about the subject, a graduate student of one of the Panel members sampled the most typical ways information is circulated--by the popular news media, both printed and electronic. This sampling was not exhaustive, although it covered newspapers and publications listed in the Readers Guide to Periodical Literature (for the period March 1972 to April 1974), and television and radio coverage (from March 1974 to May 1974). Based on this survey, the Panel concluded:

1. Both oil recovery and mineral mining in the ocean appear to be closely linked in the minds of the public--even those informed about the activities.

2. The informed public receives only limited information on ocean mining issues. In general, such information reflects the concept that benefits outweight environmental costs.

3. Although there is increasing public concern about offshore oil operations and ocean pollution, the average citizen has not been informed about many ocean mining issues.

GOVERNMENT AGENCY AND LEGISLATIVE EDUCATION

The primary educational mechanisms for government agencies and legislative bodies concerned with ocean mining have been special studies and reports, public hearings, and specialized conferences. This report is an example of one way that the Department of the Interior engages in self-education and in the promulgation of informed opinion. Although some educational and informational materials are generated for the education of government, no organization focuses directly on the preparation of educational materials concerning marine mining.

UNIVERSITY EDUCATION

With regard to formal graduate and undergraduate education at the university level, the Panel determined that each discipline of importance is covered to some extent, especially in those institutions offering interdisciplinary courses and research in ocean science and engineering. Most universities and colleges, however, are land-oriented and have few marine-oriented courses. The net result is that scientists engaged in marine minerals exploration and engineers working in marine mining require on-the-job education in ocean-related problems. Scientists and engineers trained in oceanography and ocean engineering require additional training, for example, in mineral exploration, mining systems, ore processing, and mineral economics in order to obtain the necessary professional background. Basic physics and chemistry should be included in the training. The Panel's review of marine science curricula indicates that there are at least 40 institutions offering various degrees in ocean engineering, marine technology, engineering with a marine option, and oceanography with an engineering option. But there is not one institution offering a formal degree in marine mining.[45] This may suggest an area for additional emphasis at certain universities.[46] There are however, several universities that do provide some training in marine minerals exploration and ocean mining through student participation in ocean mining research projects. These vary widely in their level of funding, degree of application, and number of student and staff participants. The majority of them are supported, at least in part, by the Department of Commerce (National Oceanic and Atmospheric Administration-Sea Grant Office) and the National Science Foundation (Office of the International Decade of Ocean Exploration). Some noteworthy of mention are:

1. University of Hawaii - research in manganese nodule distribution and geochemistry in the Pacific, and in sand mining along the insular shores;

2. Lehigh University - engineering research on sediment properties pertinent to placing stationary or mobile machines on the seafloor;

3. Columbia University (Lamont-Doherty Geological Observatory) - research projects on seafloor deposits and topography, nodule distribution, economics of marine mining, and geophysical investigations of potential nodule mining sites; and on environmental impact of deep sea mining.

4. University of California (Scripps Institution of Oceanography) - research on nodule composition in prospective mining areas of the Pacific, and seismic profiling of nodule/sediment layers;

5. University of Washington - ocean law related to seafloor mining, and nodule geochemistry;

6. Louisiana State University - studies of international law related to deep ocean mining;

7. University of Rhode Island - research on nodule chemistry, and studies in the Law of the Sea Institute;

8. University of Georgia - applied nuclear engineering research on *in situ* assessment of metal (ore) grade of nodules;

9. Texas A&M University - applied marine geology and dredging engineering;

10. University of Wisconsin - maintains a major research program devoted to seafloor minerals exploration, with ancillary projects in research and development related to outer continental shelf mining and processing systems. Some of the current projects include: exploration research on platinum, gold, tungsten, tin and rare earths in the Bering Sea outer continental shelf waters; lode copper beneath Lake Superior; lode barite beneath southeastern Alaskan waters; heavy minerals, southeastern Alaskan coast; euxinic sulfides on continental slope off Texas; nucleus/ore grade relations in deep ocean nodules; manganese pellets in Lake Michigan; hydrocyclone systems for outer continental shelf placer mining; and design of pre-mining outer continental shelf site surveys for environmental analysis. In addition, new projects are currently being planned to investigate and assess the mineral resources off certain United States insular possessions, including Samoa, Trust Territories, and Puerto Rico.

11. City University of New York (in collaboration with Lamont-Doherty Geological Observatory) - study of the environmental impact of deep-ocean mining.

12. University of Southern California - applied geological and biological research on areas of potential sand dredging in coastal California waters, and development of environmental surveys for shallow marine waters.

None of the above institutions presently offer a degree, on any level, in marine minerals exploration or ocean mining. However, the carefully guided student can, by careful selection of courses, achieve a reasonably good education in the basic sciences and engineering required by the ocean mining industry. This suggests a related educational effort in ocean mining requirements for faculty counselors in the ocean curricula. This is not likely to remain so, as the industry develops. The most direct and applicable education in academic institutions today is obtained by students who serve as laboratory and shipboard assistants in on-going, marine mining related research projects. Obviously, these opportunities are few.

Support of these projects comes largely through agency grants, notably those of the National Science Foundation-International Decade of Ocean Exploration, and the National Oceanic and Atmospheric Administration-Sea Grant. Since 1972 research grants from the industrial sector have steadily increased, largely to support graduate students and operational expenses. Some students who were associated with these academic research projects have now been employed by firms engaged in nearshore marine mining and by firms currently planning deep-ocean nodule mining. Such a system, while it provides in-depth education through research participation, cannot be expected to provide the large number of young professionals who will be needed in the next 10 years for industry and for the staffs of government agencies, including regulatory units.

TECHNICAL SUPPORT PERSONNEL EDUCATION

Marine mining requires special skills at the technician level in addition to normal shipboard activity experience. During survey, exploration, and production operations, technicians are required to operate underwater cameras and television equipment, mining machinery, instruments for nodule analysis, pilot mining tests, environmental monitoring equipment, deck machinery, tuggers, pumps and winches, as well as to perform the duties of normal seamanship.

The present educational programs, usually consisting of two years of combined classroom and shipboard instruction, can provide--with modest modification--the needed technicians. On the technician level, it is believed that on-the-job training (in addition to a two-year classroom period) will provide the kind, number and quality of technicians that will be needed.

SUMMARY

In short, public and formal education in marine mining is, like the industry, in its formative stages. Limited capabilities now exist in some university curricula, and through the international and national studies required to initiate the industry. A few young marine mining professionals are being graduated from a few universities as a result of their participation in related research projects, although no formal degrees are, as yet, being awarded. Technician training, to become effective, must have further input from the industrial and academic sectors. Expanded research support will provide technological and scientific results and specialized manpower to meet early industrial and regulatory needs. In addition, some curriculum development programs may be usefully introduced with the support of appropriate government agencies.

[45] Federal Council for Science and Technology. 1973. University Curricula in the Marine Sciences and Related Fields, Academic Year 1973-1974, 1974-1975, Revised, Washington, D.C.: Marine Technology Society.

[46] Moore, J.R. 1972. Exploitation of Ocean Mineral Resources - Perspectives and Predictions. Proceedings of the Royal Society of Edinburgh, Vol. 72, pp. 193-206.

APPENDIX A

THE PANEL ON OPERATIONAL SAFETY IN MARINE MINING

	DISCIPLINE
J. Robert Moore, Chairman Director Marine Research Laboratory The University of Wisconsin Madison, Wisconsin	Marine Geology Minerals Exploration
James M. Comstock Chief of Engineering Ocean Mining Programs Lockheed Missiles and Space Company Sunnyvale, California	Marine Engineering Marine Mining
John P. Craven Dean Marine Programs University of Hawaii Honolulu, Hawaii	Marine Engineering International Law
Marne Dubs Director Ocean Resources Department Corporate Exploration Group Kennecott-Copper Corporation New York, New York	Marine Mining Chemical Engineering
John E. Flipse President Deepsea Ventures, Incorporated Gloucester Point, Virginia	Marine Engineering Marine Mining
Don E. Kash Director Science and Public Policy Program Professor of Political Science University of Oklahoma Norman, Oklahoma	Political Science Public Policy
Martha Kohler Senior Oceanographer Scientific Development Division Environmental Services Department Bechtel Corporation San Francisco, California	Oceanography Water Resources Management

	DISCIPLINE
Oswald Roels Chairman Biological Oceanography Lamont-Doherty Geological Observatory Columbia University Palisades, New York and Professor, University Institute of Oceanography The City College of New York	Biological Oceanography Chemistry
John L. Shaw President and General Manager Ocean Management, Incorporated Bellevue, Washington	Electrical Engineering Marine Mining
Dorothy F. Soule Director Harbor Environmental Projects Allan Hancock Foundation and Adjunct Professor of Environmental Engineering The University of Southern California Los Angeles, California	Marine Biology, Ecology Environmental Engineering
Raymond Thompson Consulting Geologist Denver, Colorado	Geology Mining
Thomas M. Turner Vice President and General Manager Ellicott Machine Corporation Baltimore, Maryland	Marine Engineering Dredging

U.S. GOVERNMENT LIAISON REPRESENTATIVES

Michael Cruickshank
Conservation Division
United States Geological Survey
Menlo Park, California

William B. Gazdik
Conservation Division
United States Geological Survey
Washington, D.C.

Francis Monastero
Bureau of Land Management
United States Geological Survey
Washington, D.C.

John Padan
Pacific Marine Environmental Laboratory
National Oceanic and Atmospheric Administration
Seattle, Washington

APPENDIX B

MARINE MINING WORKSHOP PARTICIPANTS

Curtis Amuedo
Amuedo and Ivey
Denver, Colorado

Michael Baram
Massachusetts Institute of Technology
Cambridge, Massachusetts

George Brown
United States Coast Guard
Washington, D.C.

John J. Collins
American Smelting and Refining Company
New York, New York

James M. Comstock
Lockheed Missiles and Space Company
Sunnyvale, California

John P. Craven
University of Hawaii
Honolulu, Hawaii

Michael Cruickshank
U.S. Geological Survey
Menlo Park, California

George Doumani
Library of Congress
Washington, D.C.

John E. Flipse
Deepsea Ventures, Incorporated
Gloucester Point, Virginia

Richard Frank
Center for Law and Social Policy
Washington, D.C.

William Gazdik
U.S. Geological Survey
Washington, D.C.

Julian Gresser
University of Hawaii
Honolulu, Hawaii

John B. Herbich
Texas A&M University
College Station, Texas

Thomas Jennings
Bureau of Land Management
Washington, D.C.

Don E. Kash
The University of Oklahoma
Norman, Oklahoma

Martha Kohler
Bechtel Corporation
San Francisco, California

Martin Krenzke
Naval Ship Research and Development Center
Washington, D.C.

William Lee
Massachusetts Institute of Technology
Cambridge, Massachusetts

John McWilliams
Bureau of Mines
Washington, D.C.

John Mero
Ocean Resources, Incorporated
La Jolla, California

J. Robert Moore
University of Wisconsin
Madison, Wisconsin

Robert Niblock
Office of Technology Assessment
Washington, D.C.

John W. Padan
Pacific Marine Environmental Laboratory
Seattle, Washington

Oswald Roels
Lamont-Doherty Geological Observatory of Columbia University
Palisades, New York and
University Institute of Oceanography
The City College of New York

John L. Shaw
Ocean Management, Incorporated
Bellevue, Washington

Mehmet Sherif
University of Washington
Seattle, Washington

Dorothy F. Soule
University of Southern California
Los Angeles, California

Charles Sours
U.S. Geological Survey
Reston, Virginia

Paul Swatek
Massachusetts Audubon Society
Lincoln, Massachusetts

Raymond Thompson
Consulting Geologist
Denver, Colorado

Russell G. Wayland
U.S. Geological Survey
Reston, Virginia

Elmer P. Wheaton
Lockheed Missiles and Space Company (Retired)
Sunnyvale, California

Robert B. Ziegler
IHC Holland, Dredger Division
Mystic, Connecticut

APPENDIX C

OTHER CONTRIBUTORS TO THIS STUDY

Grant Ash
Corps of Engineers
Washington, D.C.

Andrew Bailey
U.S. Geological Survey
Reston, Virginia

Frederick Beck
Callahan Mining Corporation
New York, New York

Newell Booth
Naval Undersea Center
San Diego, California

George Doumani
Library of Congress
Washington, D.C.

Richard Gardner
Office of Coastal Zone Management
Washington, D.C.

Antoine Gaudin (Deceased)
Massachusetts Institute of Technology
Cambridge, Massachusetts

Amor Lane
NOAA-Department of Commerce
Rockville, Maryland

Donald Martineau
NOAA-Department of Commerce
Rockville, Maryland

Myers S. McDougal
Yale University.
New Haven, Connecticut

Edward Miles
Harvard University
Cambridge, Massachusetts

William Murden
Department of the Army
Washington, D.C.

Edward Newhouse
NOAA-Department of Commerce
Rockville, Maryland

Martin Prochnik
Department of the Interior
Washington, D.C.

Leigh Ratiner
Department of the Interior
Washington, D.C.

Eric Schneider
Environmental Protection Agency
Washington, D.C.

David Story
Senate Interior and Insular Affairs Committee
Washington, D.C.

Theodore Sudia
National Park Service
Washington, D.C.

John B. Wade
United States Coast Guard
Washington, D.C.

David Wallace
NOAA-Department of Commerce
Rockville, Maryland

George M. Watts
Coastal Engineering Research Center
Fort Belvoir, Virginia

APPENDIX D

MEMBERSHIP OF THE MARINE BOARD

ASSEMBLY OF ENGINEERING
NATIONAL RESEARCH COUNCIL

*Alfred A.H. Keil, Chairman
Dean of Engineering
Massachusetts Institute of Technology
Cambridge, Massachusetts

*Elmer P. Wheaton, Vice Chairman
Vice President and General Manager, Retired
Lockheed Missiles and Space Company
Sunnyvale, California

*Walter C. Bachman
Vice President and Chief Engineer, Retired
Gibbs and Cox, Incorporated
Short Hills, New Jersey

**Victor T. Boatwright, Jr.
Technical Assistant to the Engineering Director
Electric Boat Division
General Dynamics
Groton, Connecticut

*John P. Craven
Dean of Marine Programs
University of Hawaii
Honolulu, Hawaii

**Ira Dyer
Head
Department of Ocean Engineering
Massachusetts Institute of Technology
Cambridge, Massachusetts

**Phillip Eisenberg
Chairman
Executive Committee
Hydronautics, Incorporated
Laurel, Maryland

John E. Flipse
President
Deepsea Ventures, Incorporated
Gloucester Point, Virginia

Ronald L. Geer
Consulting Mechanical Engineer
Shell Oil Company
Houston, Texas

*Ben Clifford Gerwick, Jr.
Professor of Civil Engineering
University of California
Berkeley, California

*Earnest F. Gloyna
Dean
College of Engineering
and Joe J. King Professor
University of Texas
Austin, Texas

*Claude R. Hocott
Visiting Professor
Department of Chemical Engineering
University of Texas
Austin, Texas

*John R. Kiely
Executive Consultant
Bechtel Corporation
San Francisco, California

Christian J. Lambertsen
Director
Institute for Environmental Medicine
University of Pennsylvania Medical
Center
Philadelphia, Pennsylvania

*George F. Mechlin
Vice President, Research
General Manager, Research Laboratories
Westinghouse Electric Corporation
Pittsburgh, Pennsylvania

**J. Robert Moore
Director
Marine Research Laboratory
University of Wisconsin
Madison, Wisconsin

George C. Nickum
President
Nickum and Spaulding Associates, Incorporated
Seattle, Washington

Erman A. Pearson
Professor of Civil Engineering
University of California
Berkeley, California

**W.F. Searle, Jr.
President
Searle Consultants, Incorporated
Alexandria, Virginia

*Herman E. Sheets
Chairman and Professor
Department of Ocean Engineering
University of Rhode Island
Kingston, Rhode Island

James H. Wakelin, Jr.
President
Research Analysis Corporation
McLean, Virginia

**O.D. Waters, Jr., USN (Ret)
Professor and Head
Department of Oceanography
Florida Institute of Technology
Melbourne, Florida

*Robert L. Wiegel
Professor
Department of Civil Engineering
University of California
Berkeley, California

STAFF

Jack W. Boller
Executive Director
Marine Board
Assembly of Engineering
National Research Council

Donald L. Keach
Assistant Executive Director
Marine Board
Assembly of Engineering
National Research Council

* Member, National Academy of Engineering
** Ex-Officio Member, Marine Board

APPENDIX E

A SELECTED BIBLIOGRAPHY

Arrhenius, G. 1963. Pelagic sediments. M. N. Hill, (ed.), The Sea, New York: Interscience Publishers, pp. 655-727.

Arrhenius, G., et al. 1964. Origin of oceanic manganese minerals. Science, 144, pp. 170-173.

Baer, L. and H.L. Crutcher. 1973. Environmental predictions. I. A. Givens and A. B. Cummins, (eds.), SME Mining Engineering Handbook, Vol. 2, New York: AIME.

Barnes, Burton B. 1970. Marine phosphorite deposit delineation techniques tested on the Coronado Bank, Southern California. Offshore Technology Conference Preprints, Vol. 2, pp. 315-350.

Battelle Memorial Institute. 1971. Environmental disturbances of concern to marine mining research, a selected annotated bibliography, NOAA, ERL MMTC-3, 72 p.

Bender, M.L. 1970. Manganese nodules. R. Fairbridge, (ed.), Encyclopedia of Geochemistry and Environmental Sciences, New York: Reinhold.

Bezrukov, P. 1962. Distribution of iron-manganese nodules on the floor of the Indian Ocean. Oceanology, pp. 1014-1019.

Blissenback, E. 1972. Continental drift and metalliferous sediments. Oceanology International Proceedings, pp. 412-416.

Bonatti, E. 1972. Authigenesis of marine minerals. R. Fairbridge, (ed.), Encyclopedia of Geochemistry and Environmental Sciences, New York: Reinhold.

Brahtz, J.F. 1968. Ocean Engineering, New York: John Wiley and Sons.

Brown, B.F. 1968. Metals and corrosion. Machine Design, pp. 165-173.

Cathcart, J.B. 1968. Phosphate in the Atlantic and Gulf Coastal Plains. The Proceedings of the Fourth Forum on Geology of Industrial Minerals, pp. 23-24.

Cohn, P.D. and J.R. Welch. 1969. Power sources. J. T. Myers, (ed.), Handbook of Ocean and Underwater Engineering, New York: McGraw Hill.

Cooper, J.D. 1970. Sand and gravel. Mineral Facts and Problems, Washington, D.C.: Bureau of Mines, pp. 1185-1199.

Corp, E.L. 1970. Preliminary engineering studies to characterize the marine mining environment, Washington, D.C.: Bureau of Mines Pub. 7373.

Cronan, D.S. and J.S. Tooms. 1967. Sub-surface concentrations of manganese nodules in Pacific sediments. Deep Sea Research, Vol. 14, pp. 117-119.

Cruickshank, M.J., et al. 1968. Offshore mining: present and future. Jour. Engin. Min.

Davenport, J.M. 1971. Incentives for ocean mining: a case study of sand and gravel. Marine Technology Society Journal, Vol. 5, pp. 35-40.

Doumani, G.A. 1973. Ocean Wealth, Policy and Potential, Rochelle Park: Spartan Books.

Duane, D.B. 1969. A study of New Jersey and northern New England coastal waters. Shore and Beach, Vol. 37, No. 2, pp. 12-16.

Ebersole, W.C. 1971. Predicting disturbances to the near and offshore sedimentary regime from marine mining. Water, Air and Soil Pollution, pp. 72-88.

Eggington, W.J. and D.B. George. 1970. Application of air cushion technology to offshore drilling operations in the arctic. Offshore Technology Conference Preprints, Vol. 2, pp. 203-214.

Ehrlich, H.L. 1970. The microbiology of manganese nodules, Washington, D.C.: ONR NR 137-655, pp. 1-17.

_________. 1968. Rare Earth Abundances in Manganese Nodules, Ph.D. dissertation, Cambridge: Massachusetts Institute of Technology, 216 p.

Emery, K.O. 1960. The Sea off Southern California, New York: John Wiley and Sons.

_________. 1966. Geological methods for locating mineral deposits on the ocean floor. Exploiting the Ocean, Transcript of the Second Annual Marine Technology Society Conference and Exhibit, pp. 24-43.

_________. 1968. The continental shelf and its mineral resources. Selected Papers from the Governor's Conference on Oceanography, Albany: New York State, Department of Commerce, pp. 36-51.

_________, et al. 1970. Continental rise off eastern North America. American Association of Petroleum Geologists Bulletin, Vol. 54, pp. 44-108.

Firth, F.E., (ed.). 1969. The Encyclopedia of Marine Resources, New York: Van Nostrand, Reinhold and Company.

Hawkes, H.E. and J.S. Webb. 1964. Geochemistry in Mineral Exploration, New York: Harper and Row.

Heezen, B.C. and H.W. Menard. 1963. Topography of the deep sea floor. M. N. Hill (ed.), The Sea, Vol. 3, New York: John Wiley and Sons.

Hess, Harold D. 1971. Marine Sand and Gravel Mining Industry of the U.K., Washington, D.C.: NOAA ERL 213-MMTC 1.

Hulsemann, J. 1967. The continental margin off the Atlantic coast of the United States: carbonate in sediments, Nova Scotia to Hudson Canyon. Sedimentology Vol. 8, pp. 121-145.

Iwata, H. 1970. Research on dredging grab buckets. Proceedings of the World Dredging Conference, Tokyo.

James, H.L. 1968. Mineral resource potential of the deep ocean. Proceedings of a Symposium on Mineral Resources of the World Ocean, University of Rhode Island Pub. 4, pp. 39-44.

Jenkins, R.L. 1973. Position control. I.A. Givens and A.B. Cummins (eds.), SME Mining Engineering Handbook, Vol. 2, New York: AIME.

Kaufman, R. and W.D. Siapno. 1972. Future needs of deep ocean mineral exploration and surveying. Offshore Technology Conference Preprints, Vol. 2., pp. 309-332.

Lahman, H.S., et al. 1972. The Evolution and Utilization of Marine Mineral Resources, Cambridge: Massachusetts Institute of Technology.

LaQue, F. 1963. Materials selection for ocean engineering. J. F. Brahtz (ed.), Ocean Engineering, New York: John Wiley and Sons.

Leopold, L.B., _et al_. 1971. A procedure for evaluating environmental impact, Washington, D.C.: U.S. Geological Survey Circular 645.

Libby, F. 1969. Searching for alluvial gold deposits off Nova Scotia. _Ocean Industry_, Vol. 4, No. 1, pp. 43-47.

Maher, J.C. 1971. Geological framework and petroleum potential of the Atlantic coastal plain and continental margin, Washington, D.C.: U.S. Geological Survey Professional Paper 659.

Mauriello, L.J. and R.A. Dennis. 1968. Assessing and controlling hydraulic dredge performance. _Proceedings of the World Dredging Conference_, Rotterdam.

McIlhenny, W.F. and D.A. Ballard. 1963. The sea as a source of dissolved chemicals. _Proceedings of the 144th National American Chemical Society Meeting_.

McKelvey, V.E., _et al_. 1969. Subsea physiographic provinces and their mineral potential, Washington, D.C.: U.S. Geological Survey Circular 619.

Menard, H.W. 1964. _Marine Geology of the Pacific_, New York: McGraw Hill, 271 p.

Mero, John L. 1965. _The Mineral Resources of the Sea_, New York: Elsevier Publishing Company, 312 p.

Moore, J.R. and M.J. Cruickshank. 1973. _Identification of Technologic Gaps in Exploration of Marine Ferromanganese Deposits_, Madison: University of Wisconsin Sea Grant Advisory Report No. WIS-SG-73-404.

__________., (ed.). 1971. _Geoenvironmental and Mineral Resources_, Madison: University of Wisconsin Sea Grant Publication No. WIS-SG-71-105.

__________. 1975. Metal-bearing sediments of economic interest, coastal Bering Sea. _Symposium Proceedings of the Alaska Geological Society_, in press.

Morgan, C.L. and J.R. Moore, 1975. Role of the nucleus in formation of ferromanganese nodules: processing guidelines for the marine miner. _Offshore Technology Conference Preprints_, Vol. I , pp. 943-953.

National Commission on Material Policy. 1972. _Towards a National Materials Policy: Basic Data and Issues_, Washington, D.C.: Government Printing Office.

Owen, R.M. and J.R. Moore. 1974. Pre-mining surveys for underwater mining operations. _Proceedings of the Earth and Environment Resources Conference_, Digest of Technical Papers, Philadelphia.

______________________. 1975. Environmental Analysis of Potential Underwater Mining Sites, Madison: University of Wisconsin Sea Grant Publication No. WIS-SG-75-226.

Padelford, N.J. 1968. Public Policy and the Use of the Seas, Cambridge: MIT Sea Grant Program Publication GH-1.

Padelford, N.J. and J.E. Cook. 1971. New Dimensions of U.S. Marine Policy, Cambridge: MIT Sea Grant Program Pulbication GH-88.

Parasnis, D.S. 1966. Mining Geophysics, New York: Elsevier Publishing Company.

Ross, D.A. 1970. Atlantic continental shelf and slope of the United States: heavy minerals of the continental margin from southern Nova Scotia to northern New Jersey, Washington, D.C.: U.S. Geological Professional Paper 529-G.

Schatz, C.E. 1971. Observations of sampling and occurrence of manganese nodules. Offshore Technology Conference Preprints, Vol. 1, pp. 389-393.

Schlee, John. 1964. New Jersey offshore gravel deposits. Pit and Quarry, Vol. 57, pp. 80-81.

__________, et al. 1971. Bottom sediments on the continental shelf of the northeastern United States, Cape Cod to Cape Ann, Massachusetts, Washington, D.C.: U.S. Geological Survey Open-File Report.

Sewiorels, D.P. 1969. Stanford Engineering Analysis of Marine Resources and Technology, Stanford: Stanford University.

Sorensen, Jens C. 1971. A Framework for Identification and Control of Resource Degradation and Conflict in the Multiple Use of the Coastal Zone, Ph.D. dissertation,

Sorensen, Jens. C and W.J. Mead. 1969. A new economic appraisal of marine phosphorite deposits off the California coast, The Decade Ahead, 1970-1980, Washington, D.C.: Marine Technology Society.

Stanley, D.J., et al. 1967. Fossiliferous concretions on Georges Bank. Jour. of Sed. Petrology, Vol. 37, pp. 1070-1083.

Taney, N.E. 1971. Comments on incentives for ocean mining. Marine Technology Society Journal, Vol. 5, pp. 41-43.

Theobold, P.K., et al. 1970. Energy resources of the U.S., Washington, D.C.: U.S. Geological Survey Circular 650.

Trumbull, J.V.A., and J.C. Hathaway. 1968. Dark mineral accumulations in beach and dune sands of Cape Cod and vicinity, Washington, D.C.: U.S. Geological Survey Professional Paper 600-B.

U.S. Bureau of Mines. 1969. Minerals Yearbook: Metals, Minerals and Fuels, Volume 1-2, Washington, D.C.: Government Printing Office.

U.S. Commission on Marine Science, Engineering and Resources. 1969. Panel Reports, Volume 3, Marine Resources and Legal-Political Arrangements for their Development, Washington, D.C.: Government Printing Office.

U.S. Naval Oceanographic Office. 1970. Manned Submersible and Underwater Surveying, Washington, D.C.: U.S. Naval Oceanographic Office.

Van Baardenwijk, A.P.H. 1968. The influence of the conditions of soil on dredging output. Proceedings of the World Dredging Conference, Rotterdam.

Webb, B. 1965. Technology of sea diamond mining. Proceedings of the First Annual Marine Technology Society Conference, pp. 8-23.

Zenkevitch, N. and N.S. Skornyakova. 1961. Iron and manganese on the ocean bottom. Natura (USSR), Vol. 3, pp. 47-50.

APPENDIX F

FOREIGN CONTINENTAL SHELF DEVELOPMENTS

1. JAPAN [47]

Japan consumed 540 million metric tons (592 million tons) of aggregate for construction projects in 1971, 310 million metric tons (338 million tons) of which were natural material. Of the latter, 58 million metric tons (64 million tons), 19%, came from the seafloor. While crushed rock has begun to take the dominant position in coarse aggregate in recent years, fine aggregate has increasingly been supplied from the seafloor. In fact, 98% of the offshore production is sand--all from the near-shore areas less than 20 meters (65 ft) in water depth.

This production involves 900 small dredges of the following types:

Bucket------------53%

Sand pump---------33%

Clam shell--------14%

Environmental problems have occurred in recent years as a result of this near-shore activity. Destruction of nurseries, breakage of fishing nets, shipping casualties, and coastal erosion all have been experienced. The coastal erosion fears have become so widespread that prefecture governors have been reluctant to give new mining licenses. As a result, the Ministry of International Trade and Industry (MITI) is monitoring two test sites where it is hoped that a cause and effect relationship can be established. Unfortunately, there are as yet no biologic studies tied to this continental shelf activity.

2. UNITED KINGDOM [48]

Production of seafloor aggregate in 1970 was approximately 13 million metric tons (14 million tons), or about 13% of total United Kingdom production--a percentage that has been increasing in recent years. Except for a few old barge-mounted clam-shells, the 80 vessel United Kingdom marine mining fleet is made up of suction hopper dredges. Cargo capacities of the dredges range from about 460 to 9,200 metric tons (500 to 10,000 tons). The trend is toward larger and larger dredges to reduce the cost per unit of material dredged.

3. FRANCE [49]

Demand for construction aggregate has been growing at 11% per year in France at the same time that urban growth, zoning restrictions, and ground water problems have caused onshore resources to become less and less available. A National Center for Ocean Exploitation survey of the offshore potential has led to a four-year environmental study modeled after the United States Project NOMES (New England Offshore Mining Environmental Study) which was aborted in 1973.

Following pre-mining baseline examinations, dredging began in January 1974 at the first of two test sites. At the first site, off the mouth of the Seine, four million cubic meters (1.5×10^8 cubic ft) of sand and gravel are being removed from a rectangular pit, five meters (16 ft) deep, over a two-year period. A trailing suction hopper dredge is being used for the investigation.

The major environmental impact studies involve dispersion of the dredge-discharged silt plume, effects on benthic populations and effects on plankton.

4. EUROPE (General) [50]

Nine nations are participating in an Intergovernmental Commission on Exploitation of the Seas (ICCS) study aimed at defining the state of knowledge of the impact of marine sand and gravel mining on fisheries. In addition to the United Kingdom and France, six other European nations (as well as the United States) are considering how to respond to industrial requests for leases. Norway, Denmark, Sweden, Germany, Belgium and Ireland all are involved.

5. INDONESIA [51]

Ten to twelve bucket ladder dredges mine continental shelf tin placer deposits off Indonesia. The dredges work to the leeward of the tin islands and so enjoy the calm waters necessary for bucket ladder operations.

A recent study by the United States Geological Survey surveys the environmental problems in this area.[52]

[47] Sasaki, K. 1973. Sea Floor Sand and Gravel Mining in Japan - Present Situation and Problems, Washington, D.C.: Japan Cooperative Program in Natural Resources (UJNR) Unpublished Report, Department of Commerce.

[48] Hess, Harold D. 1971. *Marine Sand and Gravel Mining Industry of the United Kingdom*, NOAA Technical Report ERL 213-MMTC 1, Washington, D.C.: Department of Commerce.

[49] Cressard, Alain-Philippe. 1974. *Consequential Effects of Industrial Exploitation of Sand and Gravel on the Marine Environment and the Economic Activities of the Maritime Field*, Paris: Centre National pour l'Exploitation des Oceans.

[50] Personal Communication, John W. Padan, National Oceanic and Atmospheric Administration.

[51] Personal Communication, Mr. Sutedjo, Indonesian Tin Company, Bangka, Indonesia.

[52] Acuff, Dewey. 1974. *Environmental Protection Recommendations for Petroleum and Mining Operations in Asian Offshore Areas*, Washington, D.C.: U.S. Geological Survey.